I0055939

VISIR Handbook

Analog Electronics with the VISIR Remote Lab: Real Online Experiments

VISIR Handbook

Analog Electronics with the VISIR Remote Lab: Real Online Experiments

Unai Hernández-Jayo
University of Deusto, Spain

Javier García-Zubía
University of Deusto, Spain

Gustavo R. Alves
Polytechnic of Porto, Portugal

NEW JERSEY • LONDON • SINGAPORE • BEIJING • SHANGHAI • HONG KONG • TAIPEI • CHENNAI • TOKYO

Published by

World Scientific Publishing Co. Pte. Ltd.

5 Toh Tuck Link, Singapore 596224

USA office: 27 Warren Street, Suite 401-402, Hackensack, NJ 07601

UK office: 57 Shelton Street, Covent Garden, London WC2H 9HE

Library of Congress Cataloging-in-Publication Data

Names: Hernández-Jayo, Unai, author. | García-Zubía, Javier, author. | Alves, Gustavo R., author.
Title: VISIR handbook : analog electronics with the VISIR remote lab : real online experiments / Unai Hernández-Jayo, University of Deusto, Spain, Javier García-Zubía, University of Deusto, Spain, Gustavo R. Alves, Polytechnic of Porto, Portugal.
Description: New Jersey : World Scientific, [2024] | Includes bibliographical references and index.
Identifiers: LCCN 2023008552 | ISBN 9789811274145 (hardcover) | ISBN 9789811274152 (ebook) | ISBN 9789811274169 (ebook other)
Subjects: LCSH: VISIR. | Electronics laboratories--Handbooks, manuals, etc. | Electronics--Laboratory manuals. | Analog electronic systems--Study and teaching. | Analog electronic systems--Computer-assisted instruction. | Analog electronic systems--Web-based instruction.
Classification: LCC TA417.V67 H47 2024 | DDC 621.382071--dc23/eng/20230404
LC record available at https://lccn.loc.gov/2023008552

British Library Cataloguing-in-Publication Data

A catalogue record for this book is available from the British Library.

For any available supplementary material, please visit
https://www.worldscientific.com/worldscibooks/10.1142/13348#t=suppl

Desk Editors: Balasubramanian Shanmugam/Rosie Williamson/Shi Ying Koe

Typeset by Stallion Press
Email: enquiries@stallionpress.com

Preface

Remote experimentation and remote experiments have been around for more than 25 years, being commonly used in universities and other educational institutions. A remote experiment is a real experiment in which the learner is not at the same location of the experiment and the interaction is mediated by the Internet. A remote experiment is a learning tool and its main characteristics are scalability, sustainability, educational quality, and equity.

The *VISIR Handbook* focuses on one particular remote laboratory: Virtual Instruments Systems In Reality (VISIR), which is undoubtedly the best remote laboratory, the most widely used and deployed in more institutions other than the one that developed it, the Blekinge Tekniska Högskola (BTH). This book is a detailed guide to its qualities and capabilities so the reader can get to know it if he or she does not have one or exploit it in depth if he or she already has one.

VISIR is almost 25 years old now (1999–2023). It was designed by Professor Ingvar Gustavsson and it focuses on electrical and electronic circuits: assembling, measuring, and analyzing circuits created with real components and wires. From the beginning, Professor Gustavsson shared the design with the academic community. As a result, VISIR is now available in 12 countries and its design and capabilities continue to improve. This book is a compendium of everything that has been developed around the VISIR laboratory, both from a technical and didactic point of view.

The authors of this book are VISIR experts in its design and use, as well as in research around this remote laboratory. The University of Deusto (Spain) was the first to deploy a VISIR outside BTH (Sweden), while IPP/ISEP (Portugal) was the first to have two VISIRs accessible to the students. Overall, over the past 13 years, hundreds of teachers and thousands of students have used VISIR to assemble and measure several millions of circuits. It is this experience that is brought together in the book.

The reader interested in this book has three reasons or scenarios for reading and using it. The reader who is only interested in using VISIR will find in Part 2 a set of ready-to-use classroom and laboratory activities. But if the reader also wants to master VISIR as a whole so that he or she can decide which circuits to use and how to create them, then Part 1 will be perfect for his or her interests. In addition, VISIR's success and longevity allow the reader to reflect in Part 3 on various aspects of electronics education and to ask technical and didactical questions that will place him or her at the cutting edge of VISIR research. This book can therefore be read in order, but the reader can also organize his or her reading according to his or her interests.

This book considers the use of this online laboratory to support the acquisition of experimental competences. An expression of the relevance of this topic is corroborated by the recent edition of the International Handbook of Engineering Education Research (2023), which includes a chapter dedicated to "Online Laboratories in Engineering Education Research and Practice". In short, online laboratories, like VISIR, do have a role to play in better preparing engineering students for an ever-increasing online world. And even if this book focuses on remote (real) experiments for electrical and electronic circuits, a companion book also available from World Scientific Publishing entitled *Remote Laboratories: Empowering STEM Education with Technology* will provide the reader with a wider notion of how online laboratories can impact STEM education.

About the Authors

Unai Hernández-Jayo is a Telecommunications Engineer and holds a PhD in Computer and Telecommunications Engineering from the University of Deusto. He is currently a lecturer and researcher at the University of Deusto. His publications are mainly in the area of e-Learning, specifically in the design and development of remote laboratories, an activity he carries out within the framework of the DEUSTEK research group. He has participated in more than 50 research projects, being the principal investigator in more than 20 of them. He is an author/a co-author of more than 150 research publications, mainly in the area of engineering education and remote laboratories.

Javier García-Zubía is a professor at the University of Deusto since 2015 and is a member of the DEUSTEK research group. He was involved in 50+ national and international R&D projects and has authored or co-authored 250+ publications, including book chapters, conference, and journal papers with a referee process. He was invited as a keynote speaker in 10+ conferences and he was awarded in 10+ events. He edited three books around remote labs and he is the author of the book *Remote Laboratories: Empowering STEM Education with Technology* recently published by World Scientific Publishing.

Gustavo R. Alves obtained his PhD and the Habilitation in Computers and Electrical Engineering, from the University of Porto, Portugal, in 1999 and 2023, respectively. He is affiliated with the Polytechnic of Porto – School of Engineering, since 1994, where he now holds a position as an Associate Professor. He was involved in 19 national and international R&D projects, has authored or co-authored 270+ publications, including book chapters and conference and journal papers with a referee process, and has delivered 70+ invited webinars/keynotes at national and international levels. His research interests include engineering education and remote laboratories.

Dr. Alves currently serves as the Head of the Innovation Centre for Engineering and Industrial Technology (CIETI), an R&D unit supported by the Portuguese Governmental Agency for Science & Technology (FCT). He also serves as the Associate Editor for the IEEE Journal of Latin-American Learning Technologies.

Contents

Part 1

VISIR Remote Lab Description

Chapter 1

Introduction to Remote Labs in Electronics

1.1 Introduction

Let's start at the beginning... is it necessary to experiment in order to learn? And, if so, how can we develop experimental work in the teaching–learning process?

In the field of Science, Technology, Engineering, and Mathematics (STEM) education, the answer to the first question is clearly affirmative. In order to develop scientific-technological competences, students must be able to bring the theoretical concepts introduced by the teacher into the laboratory, to set up the proposed models and empirically verify their validity.

The typical place to carry out these experiments is the laboratory. It is a dynamic place where students can touch, connect, configure, move, observe, and listen; in short, configure their experiment, analyze its evolution, and reach conclusions through the analysis of the data obtained. Therefore, it is known as a hands-on laboratory.

In order to conduct a laboratory session in the hands-on laboratory, the teacher must first define the practice that the students will carry out. They must also prepare the laboratory, i.e., the equipment and materials that the students will need. Finally, they must also plan the time the students will need to complete the practical. The teacher should even have replacement equipment or materials, in case something breaks down or stops working due to a specific failure, or simply because it runs out of batteries, for example.

While it is true that these tasks are part of the teacher's work, there is also a multitude of applications, platforms, or services aimed at helping the teacher in the development of theoretical lessons; in the case of practical activities, the most common are simulators or virtual environments. In general, they are software applications that represent in a certain way the real experiments that the student could carry out in the hands-on laboratory. Simulators or virtual environments can play a very important role in the learning process, as they allow teachers to show their students variations of basic experiments that, due to lack of time, materials, and other resources (or even safety), cannot be carried out during practical sessions.

However, there are certain skills or competences that cannot be introduced via these non-real scenarios, where the responses of the experiments, given their own software-based nature, are previously defined by the developer of the simulator. That is, the responses of the experiments are always the same for the same experiment set-up and rarely give rise to failures or erroneous situations where the learner has to analyze why the results are not as expected.

Although these characteristics are not essential at basic educational levels (primary or secondary) since the aim is to bring scientific culture closer to the students, at higher levels, such as at high school and even more so at university, they are essential. At these levels, students should learn to consider concepts such as precision or accuracy in measurements or results, characteristics of a real experiment that oblige the student to employ different criteria to validate or not the starting hypothesis of the experiment.

If we focus this analysis on a specific STEM field such as electronics, a multitude of software such as Falstad, Orcad/PSpice, or Proteus, among others, is available to analyze anything from simple circuits to complex electronic systems. These tools are very helpful and widely used in the educational and professional field, but to understand how they work and why they provide the results they do, students require practical hours physically building the circuits that they will later be able to analyze using these or other programs.

The learning phase based on the manipulation, wiring, configuration, and measurement of electronic circuits is fundamental and necessary in the training process for engineers and even for beginners in secondary or high school. There is therefore a need for alternatives to simulators that could complement and assist teachers in this phase of learning, and remote labs are the technological solution.

A remote laboratory can be defined as a set of hardware and software technologies that allow a user to connect via the Internet from anywhere in the world to real equipment also located anywhere, and to carry out an experiment with almost the same performance as if they were in the hands-on laboratory. Thus, it is presented as a didactic tool that enables active learning to be carried out telematically within the framework of experimental work.

But we can also define a remote laboratory by describing precisely what it is not. A remote lab is neither a simulator nor a virtual environment. Generally speaking, in these environments, the experiment response is pre-coded or programmed by a designer and will always be the same for a given configuration. However, in a remote laboratory, the response of the experiment can also be influenced by the conditions in which the experiment is carried out, as in the hands-on laboratory: room temperature, errors introduced by the measuring equipment, tolerance of the components, etc. Thus, the user also has to analyze the response of the experiment from the point of view of the accuracy or repetitiveness of the results obtained. Concepts that move from the manual laboratory to the remote laboratory.

Although many different remote laboratories focused on various STEM disciplines can be found in the literature and are available on the Internet (García-Zubía, 2021), in this book, we will focus exclusively on the analysis of the world's best-known remote laboratory in analog electronics: the VISIR remote laboratory.

The following sections of this first chapter will contextualize the origin of, need for, and possible alternatives to the use of VISIR, emphasizing its advantages and also describing its weaknesses. Its usefulness will be analyzed with respect to what the authors have defined as the Ten Commandments of Remote Laboratories. Before doing so, however, we will first outline the structure of this book.

1.1.1 *Structure of This Book*

This book is targeted at various audiences, who may at any given time be the same: Teachers who find in the VISIR laboratory an opportunity to offer their students a teaching tool that has been widely tested and validated throughout the world (especially during the COVID pandemic); teachers who already have a version of VISIR and who do not have a reference manual to improve its operation and take advantage of its potential; or finally, researchers who can develop research work using VISIR as a technological reference or as the basis for their research.

This book is structured in three different parts:

- **Part 1 — VISIR remote part description:** First of all, the context of this book is introduced as well as the VISIR laboratory in the field of remote experimentation and the laboratories framed in the area of analog electronics in particular. The following chapter is the technical description of the architecture and fundamental elements of the VISIR laboratory. This part is especially aimed at all those who have a VISIR laboratory and need to know how to maintain it, how to add new experiments, or how to solve possible errors. It is also a part especially aimed at all those researchers who want to make contributions in the field of remote experimentation and require in-depth knowledge of the technological structure of a remote reference laboratory, such as the VISIR.
- **Part 2 — Teaching with VISIR:** This part is especially dedicated to showing real examples of how VISIR can be used in the classroom. A set of complete practices are proposed that can serve as a reference for teachers who want to include the VISIR in their portfolio of teaching tools. It also includes detailed information on how to configure the VISIR from its technological perspective to be able to deploy these experiments.
- **Part 3 — Research and reflections on VISIR:** This last part aims to summarize the history of the VISIR and provide a complete bibliography of the research work carried out about this laboratory. It is presented as a starting point to continue adding and improving the functionalities of this remote laboratory. Also in this part, the pedagogical impact of the VISIR is analyzed through the studies carried out by users during these last years.

1.2 Remote Labs in Electronics

As has already been discussed in multiple articles and references available in the literature, real experiments are indispensable in engineering teaching, as they contribute to developing skills that help understand, design, measure, and characterize physical processes. It is for this reason that laboratory experiments are integrated into the vast majority of engineering courses.

Figure 1.1. Workbench in the electronics laboratory of the Faculty of Engineering of the University of Deusto.

The workbench in an electronics laboratory is very similar in any educational center (Figure 1.1), which at the same time is practically the same as the one we can find in any company that performs electronic design. On the one hand, we have the circuit whose behavior we wish to characterize. It is generally called Circuit Under Test (CUT). Or we can have a set of components that we must assemble or connect to solve a problem and thus obtain the CUT. To perform this characterization, we have different instruments, the most common being power supply, signal generator, multimeter for current and voltage measurements, and oscilloscope. There may also be other instruments, such as a spectrum analyzer, dynamic signal analyzers, and power meters. Everything depends on the signals and data we need to perform the complete characterization of the CUT.

Based on the definition of remote laboratory introduced above, a remote laboratory in electronics must provide the possibility of designing an electronic circuit with different components and connecting different instruments to be able to characterize and measure it, all actions that the user/student must be able to carry out by means only of a browser connected to the Internet. We briefly discuss in the following different remote electronics laboratories that may (or may not) be alternatives to the VISIR.

1.2.1 *NetLab, University of South Australia*

As indicated on its website (https://netlab.unisa.edu.au/), NetLab is a remote electronics laboratory developed in response to the needs of educational innovation and support for the use of new technologies.

Regarding its architecture (Figure 1.2), NetLab is based on a dedicated server that communicates with a number of programmable laboratory instruments via IEEE 488.2 standard interface, also known as the General Purpose Interface Bus (GPIB). These instruments include the

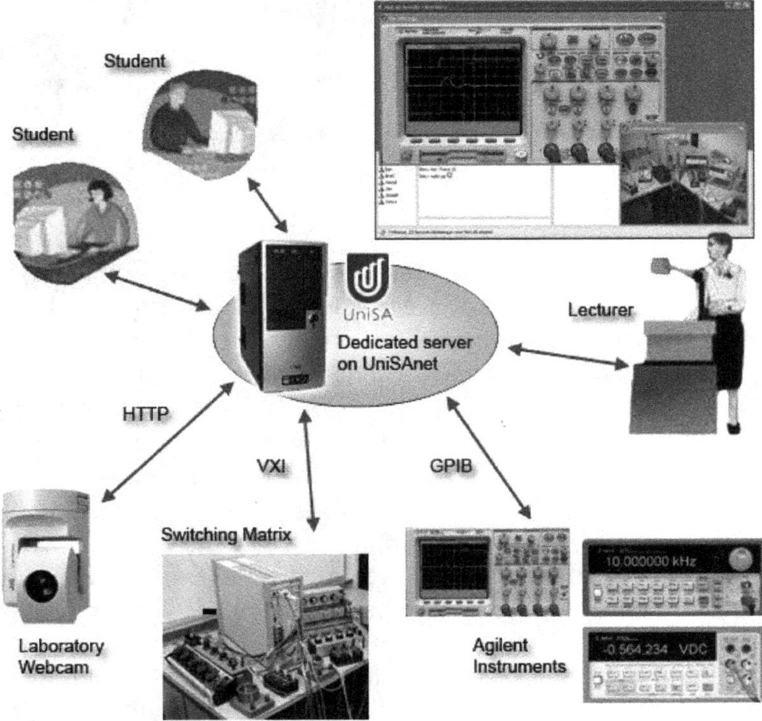

Figure 1.2. NetLab remote lab architecture.

digital oscilloscope, the function generator, and the digital multimeter. All these instruments are also connected to a 16 × 16 programmable matrix switch, which provides the user with the option of wiring and configuring various electrical circuits from available components and instruments (Nedic *et al*., 2008).

Available components are resistors, capacitors, inductors, and transformers. In addition, programmable variable resistors have been developed and interfaced into the system. Additional components can easily be added to or removed from the system at any time. NetLab also includes a camera that has its own web server and is fully controllable by the user.

To be able to carry out experiments in the laboratory, it is necessary to make a reservation beforehand. As a multiuser collaborative environment, NetLab allows more than one user to have full control of the system at the same time. However, the number of concurrent users is limited to three in order to prevent chaos in the laboratory (Nedic *et al*., 2008). Once accessed, users have at their disposal an interface composed of the interactive front panels of the laboratory instruments and the Circuit Builder (Figure 1.3). It allows electrical circuits to be wired and configured remotely. When activated, students are able to configure their own circuit required for the experiment, and then they can send their configuration to NetLab, where the real components and devices are then connected in exactly the same way via the relay matrix switch (Machotka *et al*., 2004). Therefore, according to the classification of remote laboratories proposed by García-Zubía (2021), the experiment that the user can execute is controllable, the interaction being in batch mode, and the possibility of analyzing the result in real or in deferred time, since the data generated by the

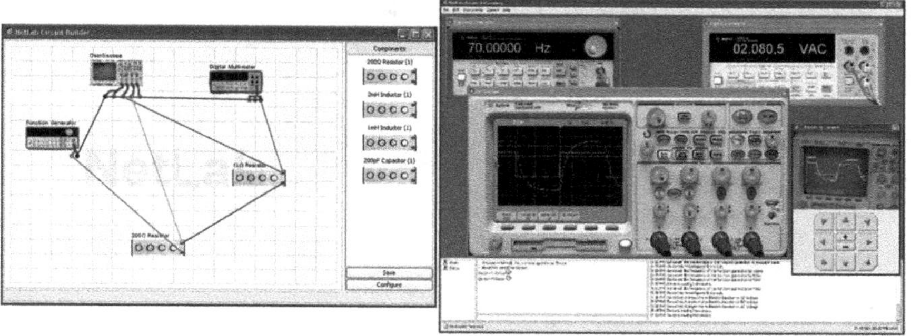

Figure 1.3. Circuit Builder (left) and example of NetLab front panel (right).

experiment can be saved and analyzed later using other programs, such as MatLab (Nafalski *et al.*, 2020).

One of the main issues with NetLab is that its software is currently implemented in Java, and some changes in Java availability have caused issues. The NetLab client was launched using Java Web Start, which is no longer supported institutionally in newer versions of Java (Nafalski *et al.*, 2020).

As for its advantages, it is worth mentioning the possibility of controlling components of variable values such as resistors, capacitors, or inductances, which offers great versatility in terms of possible circuit configurations. Moreover, the possibility of seeing in real time the execution of the experiment through a webcam offers the student a greater immersion during the laboratory session, which can also be conducted in a collaborative environment.

1.2.2 *Siddaganga Institute of Technology Remote Experiments*

Siddaganga Institute of Technology (SIT) offers its students a remote laboratory through which they can experiment with a set of pre-built electronics circuits (Figure 1.4).

Depending on the type of circuit, the user/student can perform different parameterizations of the circuit: change the voltage of the power signal and/or its frequency or modify the value of certain components among a set of pre-selected values. In no case does the user have the possibility to modify the topology of the circuit or create the circuit himself.

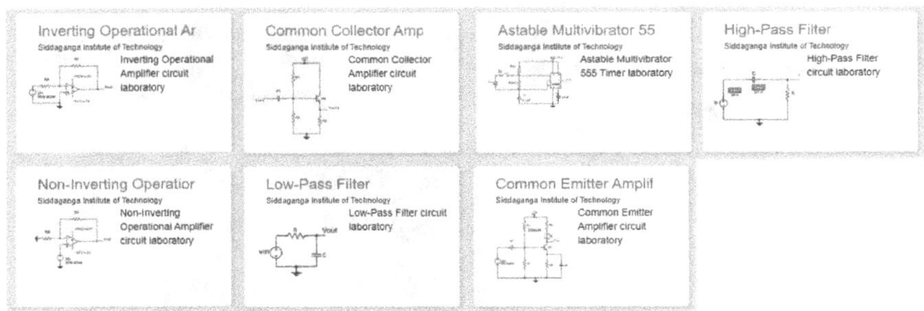

Figure 1.4. Available circuits at the Siddaganga Institute of Technology remote laboratory.

These remote labs are co-developed with LabsLand (https://labsland.com/), the global network of remote laboratories, which offers real laboratories from over a dozen countries from all the continents. Therefore, the entire software architecture of the laboratory is based on the Remote Laboratory Management System (RLMS) developed by WebLab-Deusto, the research group of which LabsLand is a spin-off (Orduña *et al.*, 2018).

The hardware structure of the laboratory is very simple. It is based on a controller that, depending on the parameterization of the circuit carried out by the user, activates the corresponding relays that allow some components or others to be connected to the circuit. As can be seen in Figure 1.5, the hardware is based on a protoboard where the LEDs indicate which component is activated at all times.

At the same time, the controller is responsible for sending the configuration orders to the instruments available in the laboratory — the oscilloscope and current meter mainly, although from the measurements obtained from the oscilloscope, the user is offered a bode diagram for analysis of the circuit. To follow the progress of the experiment, the user has a webcam that offers images of the circuit (you can see which

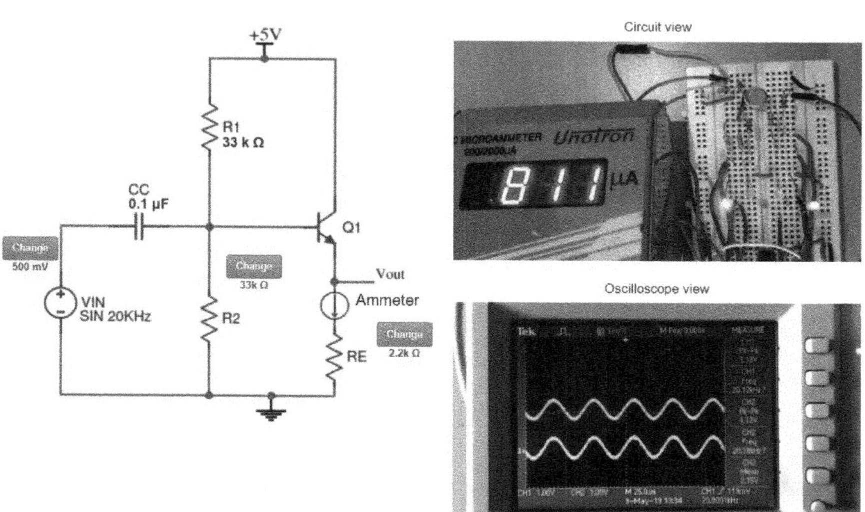

Figure 1.5. Example of common collector amplifier experiment at SIT.

component is activated at any time through the LEDs) and the oscilloscope and current meter used for the acquisition of measurements.

Again according to García-Zubía's classification (2021), it is a set of parameterizable experiments where the interaction and the results can be analyzed in real time (the user does not have the possibility of downloading the result of the experiment).

1.2.3 *RExLab, Universidade Federal de Santa Catarina*

The aim of the Remote Experimentation Laboratory (RExLab) is to allow users to access resources they do not have, and it also provides a lab environment that values experimentation as a means of developing scientific reasoning. This RLMS is known as Remote Labs Learning Environment (RELLE), which provides a series of functionalities necessary for the management of remote experiments, as well as access to the set of remote experiments provided by RExLab (Figure 1.6). In this environment, there is a wide variety of experiments in different scientific disciplines, although we are going to focus on only two of them: DC and AC electric panels.

As in the case of Siddaganga Institute of Technology, in the DC and AC laboratories provided by RExLab, the circuit topology is barely modifiable by the user. In both cases, the student can connect or disconnect certain components of the circuit, but the structure of the circuit is bounded by the panel on which it is located.

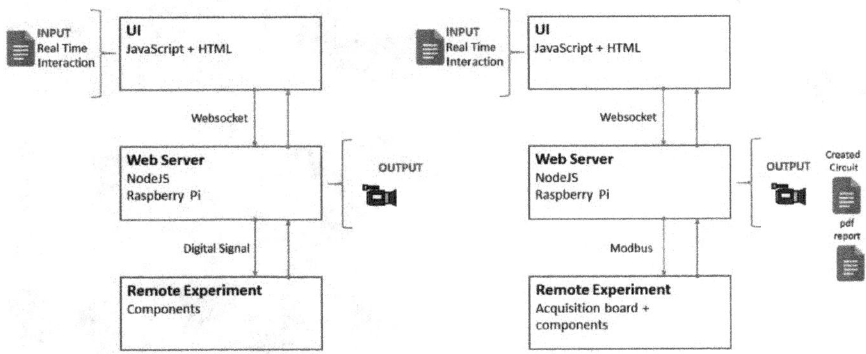

Figure 1.6. Architecture of AC electric panel (left) and DC electric panel (right).

Through the available web interfaces (Figure 1.7), the user can open or close relays that add or remove components to the CUT (resistors in the case of the DC panel and bulbs in the AC panel). As the user performs this action, represented to the right of the image obtained by the webcam is the circuit generated and being measured and observed at all times.

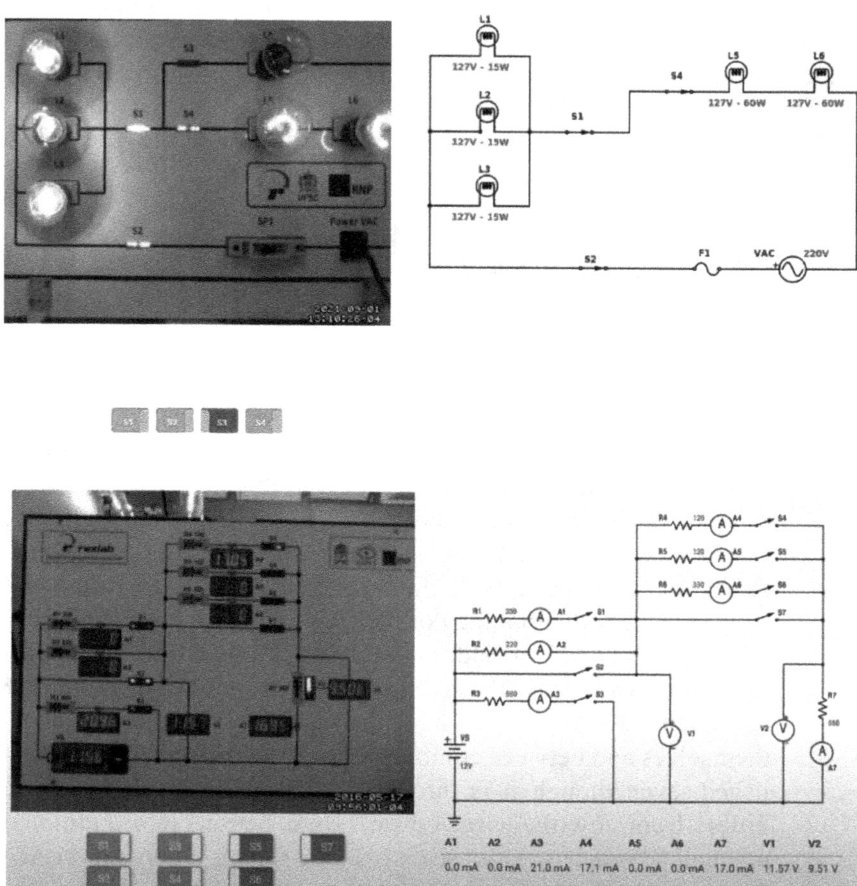

Figure 1.7. User's front panels of AC electric (top) and DC electric (bottom) remote labs.

The user can also observe the behavior of the circuit depending on the configuration made by the relays. In the case of the AC panel, the user can see how the bulbs shine more or less depending on the number of bulbs and their relative position (series or parallel). On the DC panel, the user can see the measurements offered by the ammeters and voltmeters placed on the panel.

The main limitation of these two experiments is shared: The user cannot analyze more topologies than those previously defined and established by the configuration of available relays. User configuration options are scarce, from the point of view of both circuit topology and configuration of measuring instruments.

Thus, returning to the classification established by García-Zubía (2021), these are two parameterizable experiments where the interaction and results can be observed in real time.

1.2.4 *Differences with Respect to VISIR*

The main difference and advantage that VISIR presents with respect to the remote electronics laboratories introduced in the previous sections is the versatility it offers the user with respect to the following:

- The possibility of designing the logic or structure of the experiment. This is something that is not possible in the laboratories offered by Siddaganga Institute of Technology and RExLab, which are only parameterizable, and in which the user can replace only a set of components. In the case of the NetLab laboratory, the user can define the topology of the CUT but has a limited number of components and the addition of new components to the laboratory is not easy from the point of view of maintenance. On the other hand, the way in which the connection between components themselves and between components and the instruments is established, even though it is through a high-quality web interface, differs from how the user performs these actions in a traditional manual laboratory. This distances the remote user experience from the experience you would have in a face-to-face lab.
- On the other hand, taking measurements using virtual instruments that make it possible to control their real counterparts through the web interface is only possible in both VISIR and NetLab. It is only in these laboratories that the user is able to feed the CUT

with different combinations of signals, connect the measuring instruments at different points of the CUT, and configure the instruments practically, as if they were interacting with the real instruments of the face-to-face laboratory.

- Finally, although the laboratories available at Siddaganga Institute of Technology, RExLab, and VISIR are accessible through any web browser without the need to install any specific software, in the case of the NetLab laboratory, users must download and accept the installation of a plug-in that enables them to interact with the laboratory. This may not pose a major problem for certain users, but it may reduce ease of use for many users who do not own their own PCs. Performing this installation is not easy since it requires special permission from the IT service of their institution.

This cursory analysis of the main remote laboratories of analog electronics that are currently accessible through the Internet evidences that VISIR is the one that best adapts to a real experimentation scenario. It provides the user with the same actions and benefits as if they were in the face-to-face laboratory, and they only require a connection to the network from any device and operating system, without the need for any installation or pre-configuration, which accelerates start-up and use by users.

Nonetheless, this is a research area in continuous development, so it is always interesting for the reader to keep abreast of new remote laboratories that may appear or improvements that may occur in existing ones. To this end, congresses such as the REV (Conference on Remote Engineering and Virtual Instrumentation), IEEE EDUCON (Global Engineering Education Conference), or FIE (Frontiers in Education) are events in which researchers and/or teachers publicize the new developments that may occur in this field. It was also in these congresses, among others, that the VISIR became known in the first decade of the 21st century. The following section introduces the technological evolution of this remote laboratory from its origins to the present day.

1.3 Technological Evolution of VISIR Remote Lab

The VISIR remote laboratory is the brainchild of Ingvar Gustavsson, a professor at the Blekinge Institute of Technology (BTH — Blekinge

Tekniska Högskola), a university located in southern Sweden. The first reference we can find of Professor Gustavsson's work dates from 2001. In this first work, presented at the International Conference on Engineering Education (Gustavsson, 2001), the VISIR laboratory is not yet explicitly mentioned, but one finds the first design requirements that subsequently led to its development. In this first article, Professor Gustavsson already specifies that "Physical laws must be verified by real experiments. Simulations will not do" (sic), as well as how theoretically simple it is to send the data of the instruments used to carry out the measurements over the Internet. To do this, this first version of the remote laboratory had two different deployments, both controlled from the same PC (Figure 1.8). While one of the configurations already made use of PXI (PCI Extensions for Instrumentation) technology, the other was based on GPIB (General-Purpose Instrumentation Bus).

Regarding the software architecture of this first development, it was based on the client–server paradigm, where two server programs were implemented using LabVIEW: one for the GPIB modules and another for those controlled by PXI. Meanwhile, the client was developed in Visual Basic, offering a very simple interface to access the front panels of the instruments.

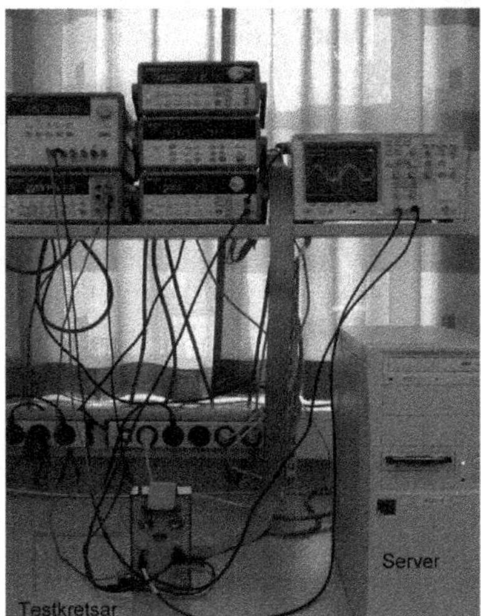

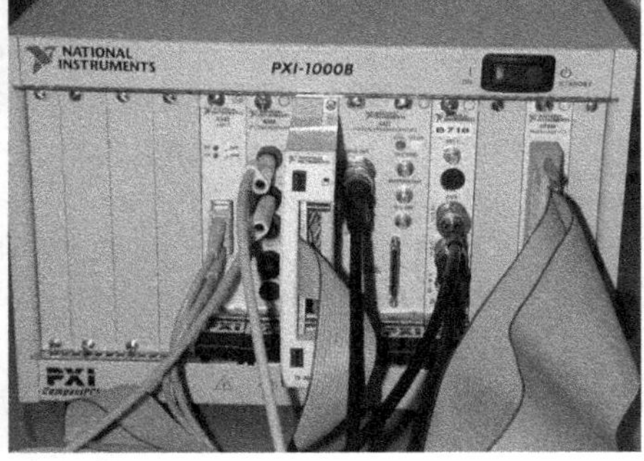

Figure 1.8. First hardware architectures of the BTH remote electronics laboratory based on GPIB technology (left) and PXI technology (right).

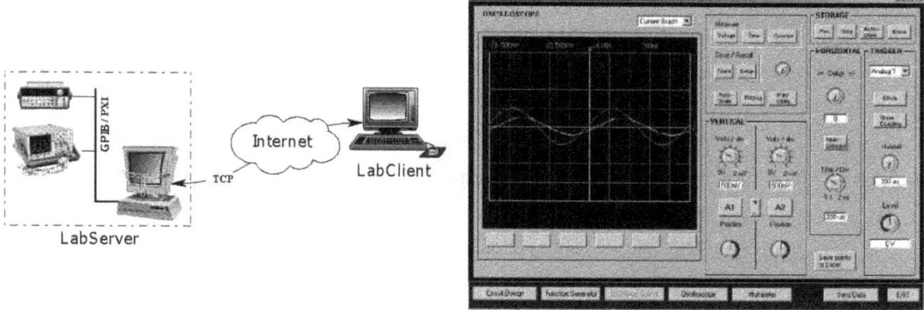

Figure 1.9. Client–server software architecture of the remote electronics laboratory (left) and client example (right).

In this first version (Figure 1.9), users had to download and install the client to be able to carry out the experiments that allowed them to verify the Kirchhoff's voltage law. This customer did not have a virtual prototype board on which the circuit could be wired, but they were already pre-built and the measurement points were also predefined.

These first tests were useful to compare the performance of both deployments and verify that the PXI configuration was the most suitable to serve several students at the same time, obtaining satisfactory response times when used with 14 students (it should be borne in mind that we are talking about the year 2001 and the students connected using a 56 kbit modem). From this first test, Professor Gustavsson already showed his interest in knowing the opinion of the users, as well as their perception as to whether the experiment was real or just a simulation, with the noise of the relays when changing their state from one circuit to another the only proof that the laboratory responded to the users' orders.

In 2002, this version of the remote laboratory was also used by the Technical University in Luleå, located more than 1000 km from the Ronneby campus of the BTH where the laboratory was installed. This was possibly the first experience of truly remote use, also offering access to the laboratory from the students' homes (Gustavsson, 2002).

Also in the year 2002, within Professor Gustavsson's framework, new remote laboratories were introduced to support subjects in the field of control systems. A webcam was added to these laboratories so that the user could observe the development of the experiment in real time. However, as Professor Gustavsson pointed out, regarding the remote electronics laboratory "*Certainly, neither video nor sound transmission from the remote lab is required because it is not possible to see electrical current with the naked eye or heard electrons moving, thereby reducing the data transfer bandwidth needed*" (Gustavsson, 2002). This statement remains valid to this day.

It is also interesting to review the satisfaction surveys that Professor Gustavsson asked his students to complete after the remote experimentation sessions. Thus, if the results compiled in his work published in 2003 in the International Journal of Engineering Education (Gustavsson, 2003a) are analyzed, general satisfaction can be observed on the part of the student users of remote laboratories. They consider the waiting times acceptable and, for example, most are of the opinion that it is possible to learn to use a laboratory equipment having "contact" with it only through the Internet. This is therefore the first evidence that the remote laboratories proposed by Professor Gustavsson were considered to be valid teaching tools.

1.3.1 *First Approach to VISIR Remote Lab: 2001–2003*

These first prototypes, as well as the positive reception by their students, contributed to the continued work on and improvement of the performance of these first remote laboratories. Focusing on the remote laboratory of analog electronics, it was also in 2003 when the first version of the switching matrix appeared, which made it possible to create circuits with 5 nodes and 10 branches, enabling students to build and measure voltages in circuits such as the one shown in Figure 1.10.

As can be seen, it is already a circuit with a certain degree of complexity and, above all, versatility because although users could not yet build the circuit in a virtual breadboard, they could select different components for each of the branches of the circuit, as seen in Figure 1.10 bottom (Gustavsson, 2003b).

To this end, a new server was developed for the remote laboratory, which appears in Figure 1.11. To the left, there is a PXI chassis (PXI-1000B) containing a controller (PXI-8176) and four plug-in boards from National Instruments. The controller comprises a PC connected to the Internet that hosts the plug-in boards in the form of two function generators (PXI-5411 and PXI-5401), an oscilloscope (PXI-5112), and a digital I/O board (PXI-6508). The instrument settings are controlled from the host computer, and there are no buttons or control knobs on the

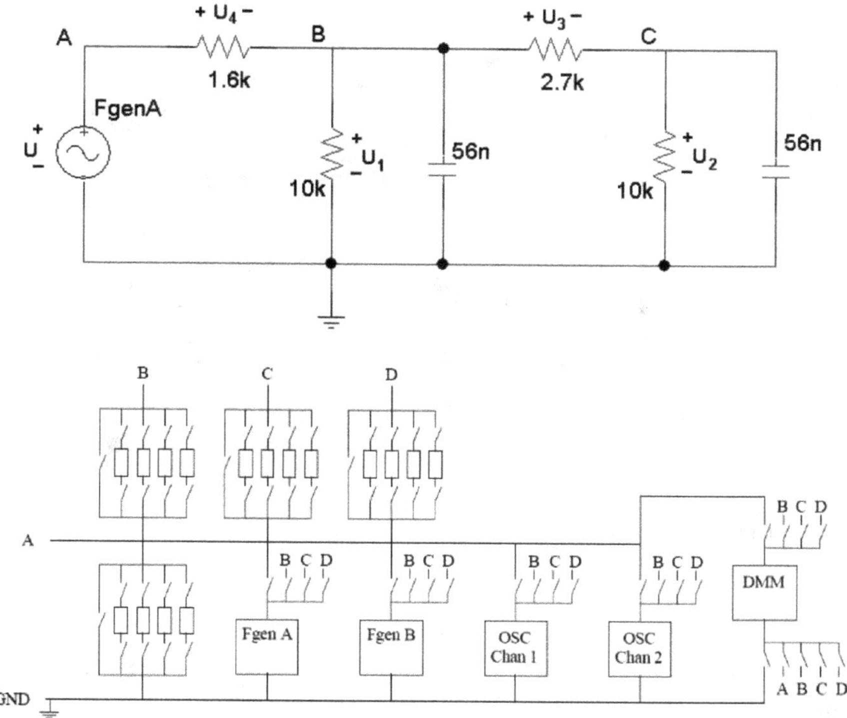

Figure 1.10. Circuit for testing Kirchhoff's voltage law in the first switching matrix (top) and example of connections to node A implemented at the switching matrix (bottom).

generators or the oscilloscope, only connectors. Next to the PXI chassis is a Data Acquisition/Switch Unit (Agilent 34970A, 34901A, 34903A, and 34904A), which functions as a multimeter. This is connected to the controller via the GPIB. To the right is a power supply (HP E3631A).

The traditional breadboard is replaced by a remotely controlled switch matrix large enough to accommodate most of the circuits used in electrical and basic electronic experiments in undergraduate education. The matrix used in the remote laboratory at BTH has five main nodes and ten main branches. The main nodes are denoted A, B, C, D, and GND. The ground terminals of the function generators and the oscilloscope are connected to GND. Each main branch can be composed of a jumper lead or up to four components with two leads mounted in parallel in holders on the printed

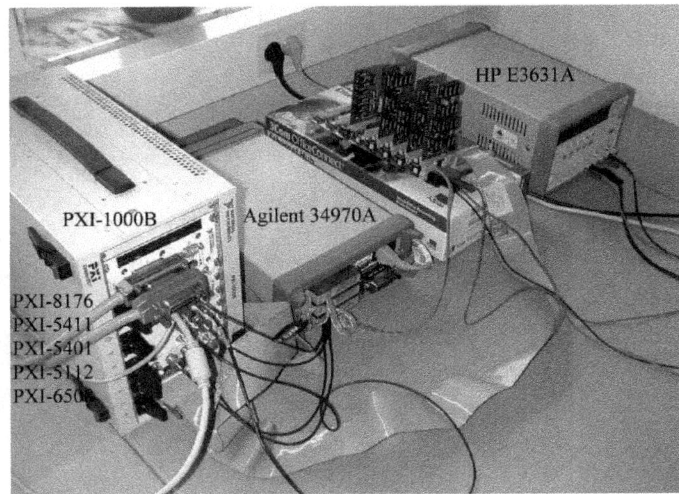

Figure 1.11. New laboratory server for experiments in circuit theory and basic electronics.

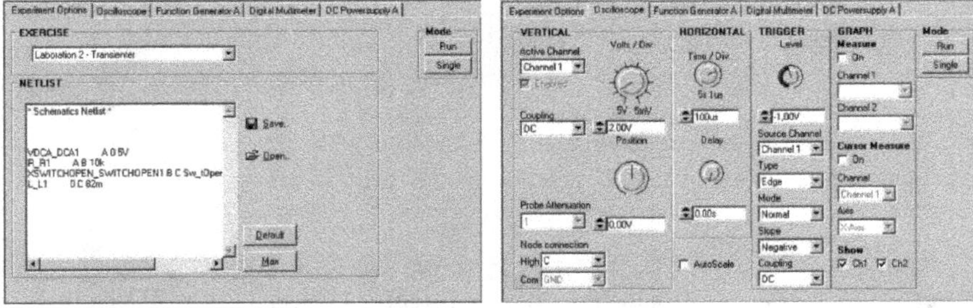

Figure 1.12. Experiment option panel (left) and virtual front panel of the oscilloscope (right).

circuit boards shown in Figure 1.11. In this way, a total of 40 different components can be connected (Gustavsson, 2003c).

In this version, the server is developed in LabVIEW 6.1 and the client is still based on Visual Basic 6, which has different tabs to access the front panels making it possible to configure the instruments or select the experiment to build on the matrix (Figure 1.12). In this version, however, the circuits are still predefined, and the user does not have the option of wiring between the components of each circuit.

1.3.2 *Second Approach to VISIR Remote Lab: 2004–2006*

It was between 2004 and 2006 that Professor Gustavsson's team began to work on a more advanced version of the remote laboratory of analog electronics. The client software was still an ActiveX control embedded in an HTML page. It was still a version that required the installation of complementary software and only allowed a maximum of eight simultaneous users (Gustavsson *et al.*, 2004).

This 2004 publication featured the first appearance of the switching matrix that uses relays to make the connections between the component boards designed expressly for the connection of the instruments in the CUT (Figure 1.13).

Another of the main elements of this laboratory is the virtual instructor, which is detailed by Professor Gustavsson in his 2005 article. The virtual instructor routine compares each desired circuit description with all the checklists and acknowledges the circuit when it matches at least one list or a subset of a list. In addition, it is responsible for verifying that the voltage and current configurations made by the user do not exceed the limits pre-established by the teacher, thus avoiding serious damage to the physical equipment of the laboratory.

This version of the remote lab is where the switching array allows complex circuits with up to 16 nodes. In this version, a component board comprises 10 sockets for components with two leads, two 20-pin IC

Figure 1.13. First version of the circuit assembly robot.

sockets for components with more pins than two, and 10 double-pole single-throw relays. This first version allowed only an oscilloscope and two function generators or a multimeter to be connected to the instrument board. As in later versions, the available nodes, identified as A–F and ground [13], are present on all boards and are interconnected with each other. It is at these nodes that the instruments and components are connected via relays to create the circuits. The laboratory technician is free to define how the components and instruments are connected. However, the ground terminals of both the oscilloscope (both channels) and the function generator are connected directly to each other and to ground to avoid problems. Also in this initial version, the function generator could only be connected to node A. This limitation was overcome in subsequent versions of the lab.

In this version, only the function generator (PXI-5401), oscilloscope (PXI-5112), and multimeter (PXI-4060) are integrated into the PXI chassis, the power supply being a standalone instrument (Agilent 3136A). There is still a maximum of eight users to whom the laboratory can respond in an optimal time, and the front panels of the instruments with which the user interacts are not yet realistic, being based on Active X (Figure 1.14).

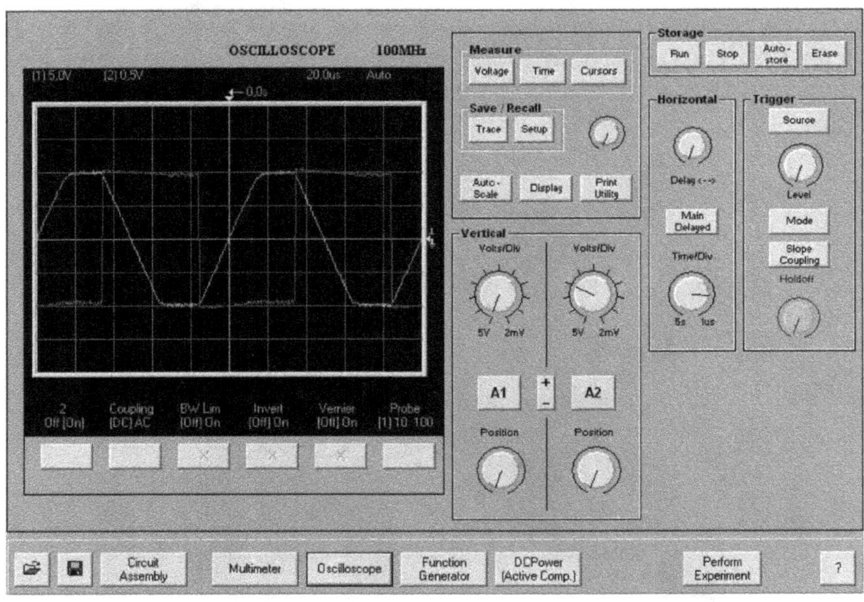

Figure 1.14. Oscilloscope virtual front panel displaying the slew rate of an uA741 operational amplifier.

1.3.3 *VISIR Remote Lab: 2006*

It was in 2006 that the acronym VISIR (Virtual Instrument Systems in Reality) was first announced, during the *9th International Conference on Engineering Education* in Puerto Rico (Gustavsson *et al*., 2006a). The last sentence of the summary of the article in which it appears is significant: "*The goal is to produce an open international standard in cooperation with universities and other organizations around the world*" (sic). A declaration of intent and the first step for the VISIR to begin to be shared, used, and installed in universities outside the BTH.

It was in the publications throughout 2006 where the interfaces of the instruments developed with Macromedia Flash first appeared, providing them with a greater realism than the previous versions and allowing greater interaction by users (Figure 1.15).

Via this web, clients can create their own circuits, connecting the components and instruments according to the teacher's specifications. Thus, the netlist created by pressing the "perform experiment" button is validated by the virtual instructor, and if it is among the list of possible netlists and the voltage and current values are among those allowed by the teacher, the circuit is sent to the server in charge of controlling the switching matrix and configuring the instruments (Gustavsson *et al*., 2006b).

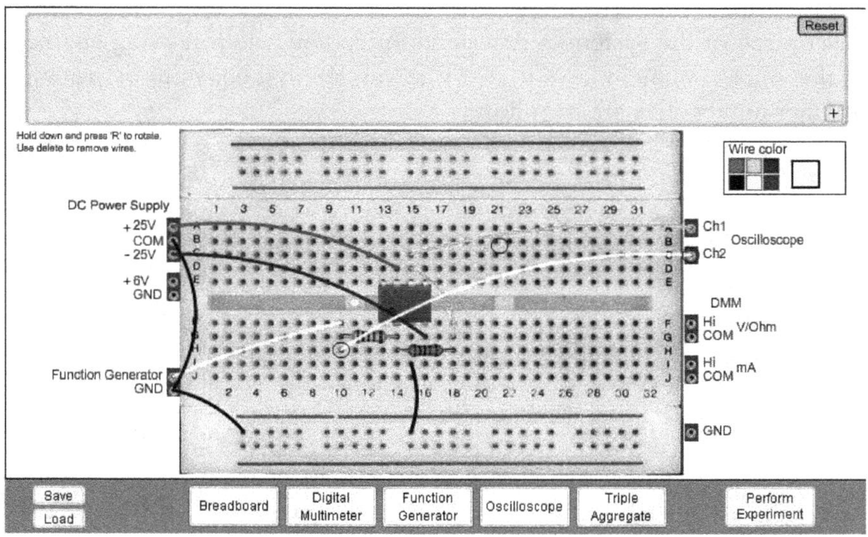

Figure 1.15. Macromedia flash-based client.

All the wiring performed by a user is taken care of in the client's computer, and the user can use as much time as they require to perform these actions. This is also true for the instrument settings. When the student presses the rightmost button Perform Experiment on the menu bar, a message containing the netlist of the desired circuit and the instrument settings is sent to the server.

This message is then decoded on the server side and placed in a FIFO (first in first out) queue. As soon as a measurement task is finished, a user request is dequeued. The virtual instructor compares the desired circuit with the max. circuits (set of available netlists defined by the teacher) to ascertain that the desired circuit is safe.

When the circuit and the instruments' set-up are verified, the power supply is set up and activated. The circuit is then created. After a 25-ms pause to allow switch transients to disappear, the instruments are set and the test probes are connected. The DMM output is read and the oscilloscope is armed. Intentional transients, if any, are fired, and the oscilloscope traces are read. The outcome is then sent back to the client's computer and is displayed on the screen. In all other cases, an error message is returned. This is the operating sequence of the remote laboratory.

It is in this article that Zackrisson *et al.* (2007) provide a complete description of the first version of the VISIR that is prepared to be used on a mass scale and offered to other institutions. Although the detailed description of the system is described throughout the following chapters of this book, the main keys to this first version that began to be installed in other universities are as follows:

- The web interface is written in PHP against a MySQL database and it is accessible through a Remote Lab Management Server (RLMS) at http://distanslabserver.its.bth.se/.
- Experiment client is written in Adobe Flash Pro 8 and contains five front panels or modules: a breadboard for wiring circuits, a function generator (representing an HP 33120A), an oscilloscope (Agilent 54622A), a triple output DC power supply (E3631A), and a digital multimeter (Fluke 23).
- These instruments' front panels can be changed and new ones can be introduced, without needing to change anything other than the instrument module.

- The measurement server is written for Microsoft Windows in C++ using Microsoft Visual C++.
- The server software is written in LabVIEW and the instrument drivers are IVI (Interchangeable Virtual Instruments)-compliant.
- The measurement request and response are transmitted using the experiment protocol, an XML-based protocol describing what settings and functions each instrument type can perform, irrespective of the hardware manufacturer.
- All instruments accessible from VISIR web interface are integrated into a PXI chassis. VISIR also is ready to control a signal analyzer (HP 35670A) through GPIB interface.
- The server is ready to serve requests from 16 simultaneous clients within less than a second.
- This version implements an authentication system based on the client's log-in. Each client is assigned with one cookie, which is generated by the web interface and stored in a database. Each user request is linked to its cookie so that the laboratory can send each user the laboratory's response by identifying its cookie.

It was therefore in 2007 that Professor Gustavsson and his team presented and offered the VISIR as an open-source software, free, functional, and stable (Figure 1.16) to all those who wanted to bring remote experimentation in the field of analog electronics to their universities.

All the necessary software, as well as the steps of the installation process, were for many years accessible and documented in http://svn.openlabs.bth.se/trac/. However, and unfortunately, since 2019 this website is out of use and is not accessible.

Regarding the hardware required to have a VISIR instance, it is necessary to acquire a PXI chassis with the necessary instruments, a LabVIEW license to execute the control of the instruments and the switching matrix (control and configuration of all of these is via the equipment server and is carried out in this environment), a computer in which to deploy the control software, and a minimum configuration of the switching matrix. Currently, it is still possible to acquire switching matrix cards through Kristian Nilsson, a disciple of Professor Gustavsson. The rest of the necessary elements can be purchased through National Instruments.

Figure 1.16. First VISIR instance at BTH at Ronneby Campus.

1.3.4 *VISIR HTML5: 2015–Nowadays*

BTH team began to share their knowledge about the VISIR platform in 2006. Ingvar Gustavsson's presence in congresses, seminars, and workshops favored the dissemination of his work, leading to the installation of new instances of the VISIR outside BTH, but this part of the story is told in Part 3 of this book.

One of the consequences of this network of VISIR laboratories that began to be created with facilities in Spain, Portugal, Austria, etc. was that not only communities of users were created but also those of developers. Research teams from universities possessing a VISIR began collaborating by adding improvements and new functions to the laboratory.

Among the most important updates made, it is worth mentioning the 2015 measure to replace the web interface developed with Flash technology with a new one implemented using HTML5 (Figure 1.17). This technological modification was necessary to be able to adapt to the requirements of the new web browsers, but mainly to make the VISIR accessible from any end with Internet access.

At the hardware level, the BTH team has also made different updates to the cards that form the switching matrix, the most notable being the

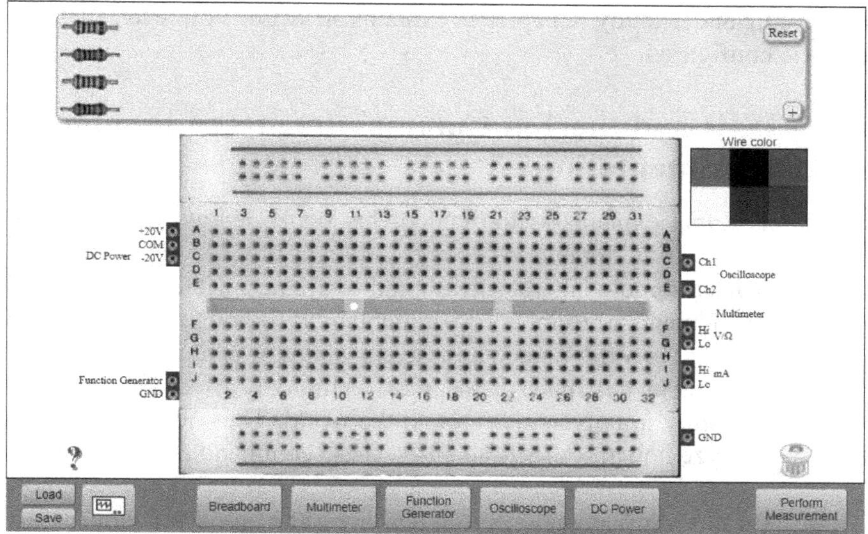

Figure 1.17. HTML5 VISIR web client.

Figure 1.18. Configurable component board versions 4.1E (left) and dual-component board version 4.1B (right).

design and implementation of the so-called dual component board. In standard component cards (Figure 1.18 left), a component is wired through its associated relay to two fixed nodes of the circuit. Meanwhile, in dual-component cards, one can have two components that may be connected to different pairs of nodes depending on the configuration needs of the circuit under study. Chapter 2 explains in detail how each of these cards is configured.

1.4 VISIR and the Ten Commandments of Remote Experimentation

As García-Zubía explains at the end of his book *Remote Laboratories: Empowering STEM Education with Technology* (2021), the design and development of remote laboratories should be framed within the so-called Ten Commandments of Remote Experimentation. Although the design and development of the VISIR predate these postulates, it is likely that its easy adaptation to them has made it the most commonly used and deployed remote laboratory in universities around the world. Thus, in this brief section, we analyze how the VISIR adapts to these commandments.

- **Do not allow designers to tell you what is good for you**: VISIR is a bare laboratory, without any educational restrictions. In other words, the teacher can and must adapt it to the requirements of the subject, making it possible for the student to perform simple or complicated experiments. It is therefore a faithful reproduction of the manual laboratory.
- **Love IT services more than your remote experiments**: The VISIR only requires a public IP to be accessed from anywhere in the world. There is no need to open special ports or have special security policies. Except for the provision of that public IP, no intervention is necessary on the part of the IT services of the institution where the VISIR is installed.
- **Be professional when choosing the remote laboratory**: From the experience of use at the University of Deusto, the remote laboratory VISIR has been used continuously since 2009, registering hundreds of thousands of accesses, with no more than a few days/hours offline, due to maintenance or fine-tuning. Also, as an example, its robustness and reliability make it the most frequently used remote laboratory offered by LabsLand, the only company in the world that offers remote experimentation services.
- **Think about the curriculum and you will succeed**: The VISIR is a laboratory adaptable to different subjects related to analog electronics, which can be used from pre-university to university courses, and which allows the development of circuits with very different levels of difficulty.

- **The curriculum is the teacher's, not the designer's**: The VISIR is a laboratory that can be used as a teaching tool and that contributes to the development of general and specific competences. It is the teacher who can adapt the tool to the curriculum by proposing different activities and experiments.
- **Control the tool and not vice versa**: At the VISIR remote lab, it is the teacher who decides the complexity of the circuits to be carried out. They determine the number and type of components to be used in each experiment, as well as the instruments to be used. The teacher fully controls the characteristics and functionalities of the laboratory made available to their students.
- **Protect your students from your sense of innovation**: The teacher needs a relatively short time to learn how to use the VISIR. From that moment on, they can bring the real electronics lab into the classroom via the Internet and combine it with other digital tools: online quizzes, shared spreadsheets, etc. It is the teacher who must give a pedagogical sense to all of this so that the student understands what is the starting point, the need and reason for using the VISIR, and the learning objective.
- **Students are not lab rats**: VISIR is a tool mature enough and sufficiently widely used by teachers throughout the world to ensure that it is a valid educational tool that favors the student's learning process.
- **The experiment should help, it should not in itself be a challenge**: VISIR has always been conceived of as a means, not an end in itself. That is why its graphical interfaces are practically the same as those that students can find in the face-to-face laboratory, which speeds up the learning process. In this way, students' user experience is very similar to what they can find in a face-to-face laboratory, and they can even avoid recurring problems, such as protoboards breaking or cables not making good contact — problems that frustrate students and do not arise in the VISIR.
- **Students are digital natives, all of them!**: The level of interaction and realism of the VISIR is very high, which makes students acquire the necessary skills very quickly. By having access to real instruments through virtual panels, they lose the fear of "breaking something" and are more likely to experiment with circuits and equipment that can also control in the face-to-face laboratory.

References

Carr, S., Machotka, J., & Nedic, Z. S. (2006). Real, remote and virtual laboratories. In *Proceedings of UNESCO International Centre for Engineering Education UICEE 8th Annual Conference on Engineering Education*, Kingston, Jamaica, pp. 223–226.

García-Zubía, J. (2021). *Remote Laboratories. Empowering STEM Education with Technology* (p. 67). London: World Scientific Publishing.

Gustavsson, I. (2001). Laboratory Experiments In Distance Learning. *International Conference on Engineering Education, Session 8B1, August 6–10, Oslo, Norway.*

Gustavsson, I. (2002). Remote laboratory experiments in electrical engineering education. *Proceedings of the 4th IEEE International Caracas Conference on Devices, Circuits and Systems* (pp. I025-1, I025-5).

Gustavsson, I. (2003a). A remote access laboratory for electrical circuit experiments. *International Journal of Engineering Education*, 19(3), 409–419.

Gustavsson, I. (2003b). Traditional laboratory exercises by remote experimentations in electrical engineering education. In *INNOVATIONS 2004: World Innovations in Engineering Education and Research, Arlington: iNEER* (pp. 163–172).

Gustavsson, I. (2003c). User-defined electrical experiments in a remote laboratory. *In Proceedings of the 2003 American Society for Engineering Education Annual Conference & Exposition* (pp. 8.1233.1–8.1233.10).

Gustavsson, I., Zackrisson, J., & Olsson, T. (2004). Traditional lab sessions in a remote laboratory for circuit analysis. Paper Presented at the *15th EAEEIE Annual Conference on Innovation in Education for Electrical and Information Engineering*, 2004, Published.

Gustavsson, I., Olsson, T., Åkesson, H., Zackrisson, J., & Håkansson, L. (2005). A remote electronics laboratory for physical experiments using virtual breadboards. Paper Presented at the *ASEE Annual Conference*, 2005, Published.

Gustavsson, I., Zackrisson, J., Ström Bartunek, J., Åkesson, H., Håkansson, L., & Lagö, T. (2006a). An Instructional Electronics Laboratory Opened for Remote Operation and Control. In *Proceedings of the 9th International Conference on Engineering Education* (pp. M4G-1–M4G-6).

Gustavsson, I., Zackrisson, J., Åkesson, H., Håkansson, L., Claesson, I., & Lagö, T. (2006b). Remote operation and control of traditional laboratory equipment. *International Journal of Online Engineering*, 2(1), 1–8.

Nafalski, A., Milosz, M., Considine, H., & Nedić, Z. (2020). Overseas use of the remote laboratory NetLab. 2020 *IEEE Global Engineering Education Conference (EDUCON)*, Porto, Portugal, pp. 568–573. doi: 10.1109/EDUCON45650.2020.9125230.

Nedic, Z., Machotka, J., & Nafalski, A. (2008). Remote laboratory Netlab for effective interaction with real equipment over the internet. *2008 Conference on Human System Interactions*, Krakow, Poland, pp. 846–851. doi: 10.1109/HSI.2008.4581553.

Orduña, P., Garcia-Zubia, J., Rodriguez-Gil, L., Angulo, I., Hernandez-Jayo, U., Dziabenko, O., & López-de-Ipiña, D. (2018). The WebLab-Deusto remote laboratory management system architecture: Achieving scalability, interoperability, and federation of remote experimentation. In M. Auer, A. Azad, A. Edwards & T. de Jong (Eds.), *Cyber-Physical Laboratories in Engineering and Science Education* (pp. 17–42). Springer.

Zackrisson, J., Gustavsson, I., & Håkansson, L. (2007). An overview of the VISIR open source software distribution 2007. In *Proceedings of the 2007 International Conference on Remote Engineering and Virtual Instrumentation (REV)* (pp. 1–14).

Chapter 2

VISIR Remote Lab

2.1 Introduction

According to the characteristics defined by García-Zubía (2021), the VISIR laboratory, given the nature of the experiments that can be performed there, is a real remote laboratory, since the circuits designed by the user in the web interface are built in a real way in the switching matrix that is part of the laboratory's hardware infrastructure. Regarding the user's interaction with VISIR, the laboratory executes the configuration of the circuit and the taking of measurements in batch mode, although it is true that to carry out these actions, only a few milliseconds are necessary. User perception is that of running an interactive session and obtaining the data in real time.

VISIR offers users full control of the experiment since they can change both the logic structure of the experiment (build the circuit they want) and the excitation values of the input signals. Similarly, a user can configure the measuring instruments on a variety of scales, making it possible to analyze the behavior of the circuit under test from different perspectives. These are therefore combinational experiments in which, depending on the input parameters, the user can observe different outputs.

We must not forget that the VISIR remote laboratory is designed to be used within the didactics and practice of analog electronics. This does not preclude its use for the analysis of digital circuits (an example is given in Part 2 of this book), although the capacity offered by the laboratory will not be as powerful.

Operation of the laboratory, from the user's perspective, is simple: Through the web client, users have access to a prototype board (breadboard) where they can do the following actions:

(a) Access a set of components that were previously selected by the teacher or the entire range of components available in the laboratory. As in a traditional laboratory, when the student attends a laboratory session the teacher offers the components to use during the experiment, or the user has full access to all the components available in the lab. We will see later that in VISIR, the library/cabinet of available components is defined by the file `components.list`.
(b) Interconnect the components to create a circuit. The connections the user can make must be previously defined by the laboratory technician, the person in charge of setting up the laboratory according to the teacher's needs. These connections, which will be introduced later, will be defined by the `.max` files.
(c) Power the circuit. For this, the user can set up a Direct Current (DC) power supply that has three different voltage outputs (+25 VDC, –25 VDC, or 6 VDC) or a function generator that allows users to feed the circuit under test with periodic signals with different amplitudes, frequencies, and waveforms.
(d) Measure circuit variables using either a digital multimeter or an oscilloscope. The multimeter allows the user to take voltage and current measurements of DC and AC signals, as well as resistance values. The oscilloscope makes it possible to visualize and measure signals.

It is also important to differentiate between the different roles involved in the process of starting up and executing an experiment in VISIR. The roles to consider are as follows:

- **User**: This is the role played by the person who interacts with the VISIR through its web interface/client. On some occasions, we can refer to the user as a student because in most cases, students will be the main users of the VISIR.
- **Teacher/instructor**: This is the person who uses the VISIR in their classes to introduce concepts of electronics to their students. The teacher, based on these concepts, determines what components and experiments they wish to make available to their students during the practical classes. For this reason, the teacher must have fluid communication with the laboratory technician.

The teacher does not require in-depth knowledge of the hardware and software architecture of the VISIR. However, having the basic notions will allow them to know in advance the capacity of the laboratory and the experiments that can or cannot be performed.

- **Laboratory technician**: This is the person in charge of the maintenance of the VISIR, their main task being to set up the laboratory according to the experiments the teacher wants to perform during the classes. That is why technicians must be familiar with both the hardware and software structures of the VISIR. In many cases, the same person can be a teacher and a technician, but the recommendation is that there is only one technician in charge of configuring the laboratory so that no information is lost and experiments are updated in an orderly manner.
- **Institution**: This plays a fundamental role. The deployment and use of the VISIR by teachers must be supported by the department. The VISIR has demonstrated its usefulness as a teaching tool and the characteristics it offers in the development of competences, so the institution that hosts it must also offer facilities for its installation (IT services) and for its use (integration within educational programs).

The following sections provide a technological description of both the software and hardware architecture of the VISIR. These sections are of particular interest not only to the laboratory technician in charge of maintaining the VISIR but also to teachers, so they are familiar with the possibilities offered by the laboratory.

This chapter ends with a section dedicated to the configuration of the laboratory, written as a guide to the installation and maintenance of the VISIR. This is therefore a reference tool to update experiments, and a technological introduction that makes it possible to establish a starting point for future research works based on this remote laboratory.

2.2 VISIR Software Architecture

VISIR software architecture is divided into three distinct parts: An Experiment Client (EC), through which users can control the experiments; the Measurement Server (MS), responsible for handling experiment requests from the ECs; and finally, the Equipment Server (EqS), which is

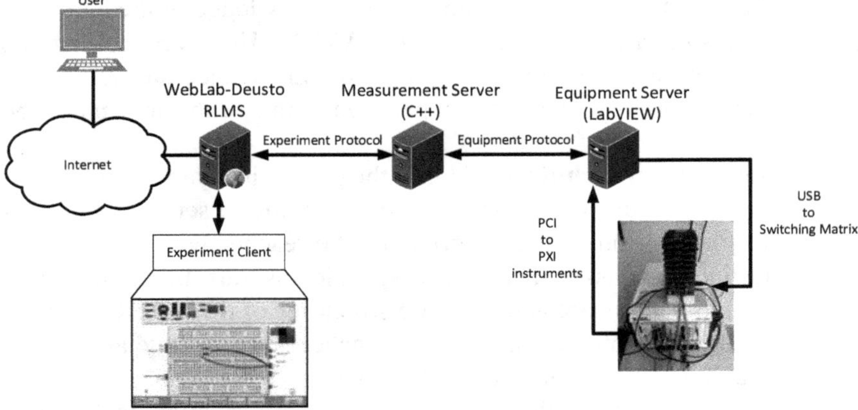

Figure 2.1. VISIR software architecture.

a standalone equipment controller handling the low-level instrument interfaces (Figure 2.1).

These are the three software elements that allow the user to create a circuit using a web application, ensure that it meets the safety and design specifications determined by the teacher, and finally, build and measure it using the switching matrix and instruments contained in the laboratory. However, these three elements must be "wrapped" by a Remote Lab Management System (RLMS). The RLMS suggested by the authors of this book is the one offered as open source by the WebLab-Deusto group (http://weblab.deusto.es/). It is a system that facilitates management of remote laboratories regardless of their nature and type.

Every remote laboratory must manage features such as authentication, authorization, scheduling users to ensure exclusive accesses (typically through a queue or calendar-based booking), user tracking, and other administration tools. These features are common to most remote laboratories and are independent of the remote laboratory settings. All these functionalities are offered by the WebLab-Deusto RLMS. Moreover, this RLMS provides development toolkits for developing new remote laboratories. The main idea is that by adding a new feature to the RLMS (e.g., supporting Learning Management System — LMS), all the laboratories developed on top of these systems will support this feature automatically. In fact, the remote laboratory, thanks to the features of this RLMS, can be accessed from different LMSs, such as Moodle.

The description of this RLMS is beyond the scope of this book, but more information, documentation, and its software can be downloaded for free from https://github.com/weblabdeusto/weblablib. Returning to the VISIR, let us describe its three main software elements.

2.2.1 *VISIR Experiment Client*

When an experiment session is successfully started via the RLMS, the system presents the user with the means available for experimentation, i.e., the EC software. This is achieved via a webpage built in HTML5 that contains the EC, i.e., the VISIR client. Parameters to control the set-up of the experiment can be given to the EC, for instance, which instruments to display or which components to show by default on the breadboard. Course teachers can control these parameters. The web interface itself adds some parameters that cannot be controlled, e.g., which MS to contact.

In addition, a special cookie parameter identifying the initiated experiment session is given to the EC. Each client logged into the system is then identified by this cookie. In this way, each response generated by the laboratory to each request by each client is identified by this cookie. Thus, the remote laboratory can manage multiple connected users and offer each of them the individual response generated by the laboratory.

The main element of the EC is the interface through which the user can build the circuit and set up the instruments to excite/feed and perform measurements. To this end, a life-like experiment environment has been developed, containing the set of instruments included in VISIR: breadboard, function generator, power supply, oscilloscope, and multimeter (Figure 2.2). Each of these instruments is represented by a virtual front panel, which represents an instrument usually found in a classic laboratory, presenting all the usual functions of the hardware instrument. Photographs of the instruments are used further to improve the perception of the instrument as real.

Thanks to this set-up, students will familiarize themselves with classic instruments by using the ones in the remote laboratory. This means that the skills they develop using these virtual front panels are the same or valid when they are in front of the real instruments in the hands-on lab. Each instrument is contained in its own module, which can be loaded on demand when the client starts. In fact, it is possible to have a different

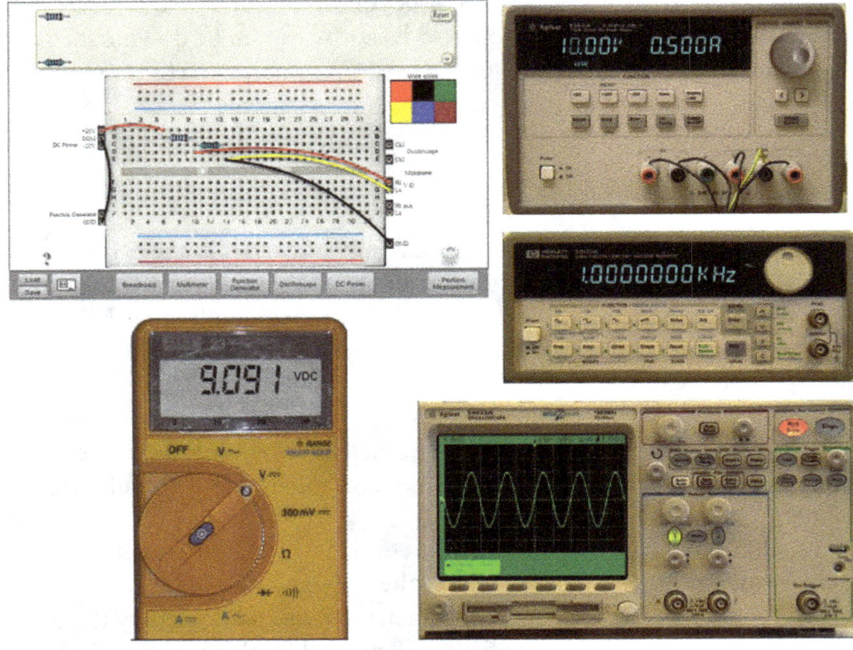

Figure 2.2. Available instruments from panels at the VISIR remote lab.

front panel for each instrument, for example, representing instruments from different manufacturers. Which instruments to load is decided by the parameters given to the EC by the web interface. The modular approach has the advantage of being very flexible when it comes to changes. Instruments can change and new ones can be introduced, without the need to change anything other than the instrument module, i.e., its front panel. Current modules available are as follows:

- a breadboard for wiring circuits
- function generator, HP 33120A
- oscilloscope, Agilent 54622A
- triple output DC power supply, E3631A
- digital multimeter, Fluke 23
- signal analyser, HP 35670A.

The main tasks of the EC are to allow users to build their own circuit, power the circuit using the function generator (AC circuits) or the power supply (DC circuits), perform measurements, and view these results through the digital multimeter or the oscilloscope front panels. When a user makes a measurement, the EC takes the settings from all the instrument modules and compiles them into a measurement request that is formatted according to the Experiment Protocol (Figure 2.3). This request is subsequently sent to the MS, which returns a response that can be read by the instruments to update their displays. The measurement request and response are transmitted using the Experiment Protocol, an XML-based protocol describing the potential settings and functions each instrument type can perform, irrespective of the hardware manufacturer. This makes it possible to implement new modules, for example, emulating an instrument that is not available in the current instrument set.

2.2.2 *Experiment Protocol*

Experiment Protocol defines the rules for the exchange of information between the EC and the MS. This protocol, created specifically for the VISIR laboratory, is based on an XML file and a request–response operating structure. It is the EC that makes a request for configuration of the laboratory instruments, and it is the MS that responds. There are two types of MS responses:

(1) If the virtual instructor satisfactorily resolves the EC request, this means the circuit can be built and measured in the laboratory. In this case, the MS sends a request to the EqS following the rules established by the Equipment Protocol. The EqS performs the following actions: (a) configure the switching matrix with the user-defined circuit, (b) set up the instruments to power the circuit (DC power supply and/or the function generator), and (c) set up the instruments (multimeter and/or oscilloscope) to take measurements. Once the measurements have been taken, the EqS sends the response to the MS according to the format defined by the Equipment Protocol. Finally, the EC shows the measurements taken to the user, setting the virtual panels of the EC instruments according to the Experiment Protocol information.

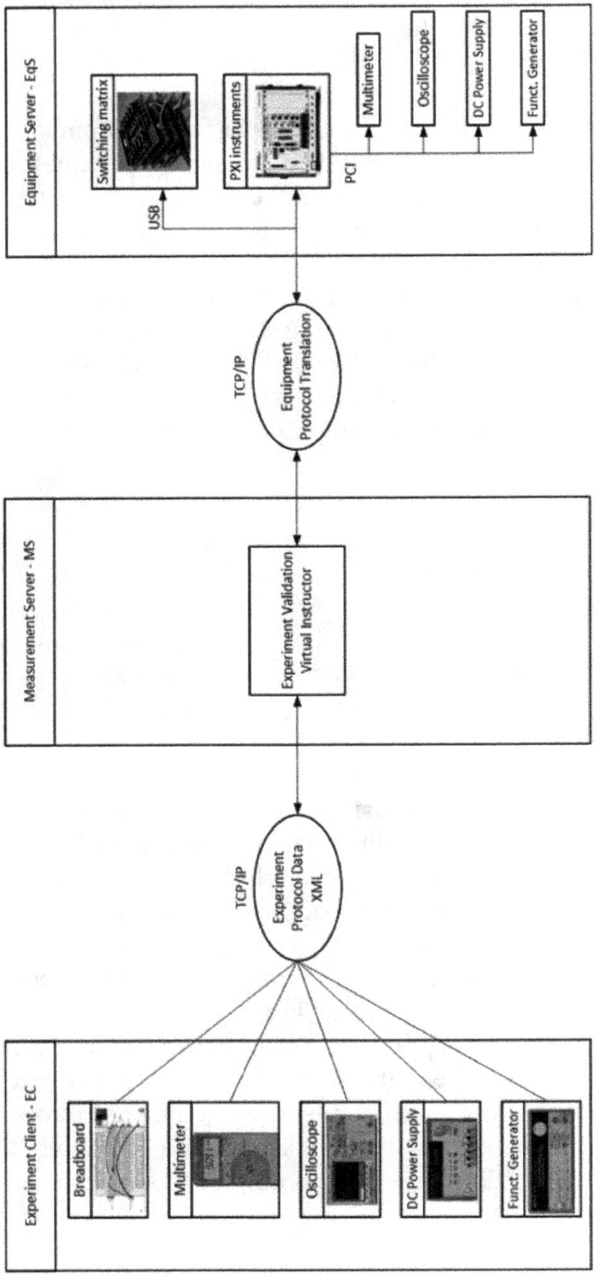

Figure 2.3. Instrument data transmitted from experiment client to hardware instruments.

The circuit cannot be constructed. Either it is unsafe or the current set of rules validating the circuit can't find a suitable solution.

Figure 2.4. Error message reported by VISIR.

(2) If the virtual instructor cannot resolve the circuit, because it is dangerous or because the combination of components and connections has not been previously validated by the teacher and the laboratory technician, the MS returns an error message to the EC (Figure 2.4).

The operation of the virtual instructor is described in more detail in the section corresponding to analysis of the MS (Section 2.2.3).

As indicated above, the Experiment Protocol is a file with a structure based on the XML language that defines the requests sent by the EC to the MS and the responses sent from the MS to the EC. In the following, we introduce each of the parts of which there is an example of a file containing a request from the EC to the MS.

It is important to note that regardless of whether the student uses or configures all the instruments, the XML file that is exchanged during the request and response operations has the same structure. If an instrument is not configured, its default parameters are sent.

2.2.2.1 *Circuit under test description — request*

Both the file containing an EC request and the file containing the MS response begin with the tags describing the version of the protocol being used, as well as the unique cookie that identifies the user making the request. This cookie, identified as session key, is the ID that allows the VISIR to return to each user the response corresponding to their request (Code 2.1).

Following these tags, the description of the circuit the user has built on the breadboard appears. This description identifies the components

```
<protocol version="1.3">
<request sessionkey="79b9674e14a9c1005fa4c1d3cd723aa0">
```

Code 2.1. Description of Experiment Protocol version and session key at request frame.

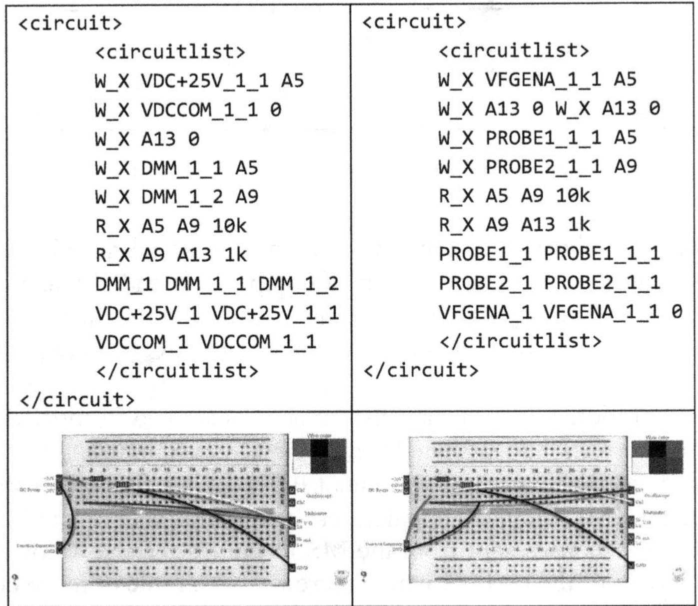

```
<circuit>
	<circuitlist>
	W_X VDC+25V_1_1 A5
	W_X VDCCOM_1_1 0
	W_X A13 0
	W_X DMM_1_1 A5
	W_X DMM_1_2 A9
	R_X A5 A9 10k
	R_X A9 A13 1k
	DMM_1 DMM_1_1 DMM_1_2
	VDC+25V_1 VDC+25V_1_1
	VDCCOM_1 VDCCOM_1_1
	</circuitlist>
</circuit>
```

```
<circuit>
	<circuitlist>
	W_X VFGENA_1_1 A5
	W_X A13 0 W_X A13 0
	W_X PROBE1_1_1 A5
	W_X PROBE2_1_1 A9
	R_X A5 A9 10k
	R_X A9 A13 1k
	PROBE1_1 PROBE1_1_1
	PROBE2_1 PROBE2_1_1
	VFGENA_1 VFGENA_1_1 0
	</circuitlist>
</circuit>
```

Code 2.2. Circuit under test description at the Experiment Protocol.

used, the connections made between the components, and the connection of the instruments with the circuit.

Code 2.2 shows an example of two circuits built on the breadboard. On the left, a circuit formed by two resistors of 10 kΩ and 1 kΩ connected in series is characterized. The circuit is fed with the direct current source of +25 VDC and the multimeter is used to measure the voltage drop in the resistance of 10 kΩ. One can see how the identifier W_X describes the wires, R_X identifies where the resistors are connected (if, for example, there were capacitors, the identifier would be C_X), as well as specific identifiers for each of the instruments used and the connections that exit each of them.

As an example, it is indicated that the multimeter is identified by the code DMM_1 (if two multimeters are available, the second is identified as DMM_2) and has two connectors, identified as DMM_1_1 and DMM_1_2. The DMM_1_1 (the wire from HI channel of the DMM to the resistor) test point is connected to the connection point A5, to which the output of +25 VDC (VDC + 25 V_1_1 connection) and the left terminal of the

10 kΩ resistor (R_X A5 A9 10k) are also connected. The right-hand terminal of this resistor is connected to point A9 of the circuit.

Also apparent is the fact that in the breadboard, all connection points are connected in columns and identified from A1 to A32 at the top and from B1 to B32 at the bottom.

Thus, the example on the right presents the description of the circuit when the function generator is used to power the circuit and the oscilloscope to observe the signals that represent the voltage at the two points of the circuit to which the CH1 and CH2 channels of the oscilloscope have been connected, i.e., A5 and A9.

2.2.2.2 *Multimeter description — request*

The description of the multimeter is quite simple, as it only contains the values indicated in Code 2.3 and only the parameter dmm_function can be set up by the user by turning the selector of the measurement on the virtual panel. In these examples, first, a DC voltage measurement is selected by the user, and then the resistance between both terminals of the multimeter is measured. The other parameters are always set up by default with the same values. The identification parameter id="1" is necessary when more than one multimeter is available in VISIR.

```
<multimeter id="1">
        <dmm_function value="dc volts"/>
        <dmm_resolution value="3.5"/>
        <dmm_range value="-1"/>
        <dmm_autozero value="1"/>
</multimeter>
```

```
<multimeter id="1">
        <dmm_function value="resistance"/>
        <dmm_resolution value="3.5"/>
        <dmm_range value="-1"/>
        <dmm_autozero value="1"/>
</multimeter>
```

Code 2.3. Multimeter description in the Experiment Protocol.

2.2.2.3 *Function generator description — request*

After the description of the multimeter set-up, the configuration parameters of the function generator appear. In this instrument, four parameters

```
<functiongenerator id="1">
        <fg_waveform value="sine"/>
        <fg_frequency value="1000"/>
        <fg_amplitude value="0.5"/>
        <fg_offset value="0"/>
</functiongenerator>
```

```
<functiongenerator id="1">
        <fg_waveform value="square"/>
        <fg_frequency value="1000000"/>
        <fg_amplitude value="2"/>
        <fg_offset value="0.5"/>
</functiongenerator>
```

Code 2.4. Function generator description in the Experiment Protocol.

can be set up: waveform (sine, square, triangle, and ramp-up), frequency (from 1 Hz to 1 MHz), amplitude (maximum value is 10 Vpp), and offset (this value depends on the amplitude set-up). Code 2.4 includes two examples of set-up performed by the user at the EC and requested from the MS.

2.2.2.4 *Oscilloscope description — request*

A description of the oscilloscope set-up is included in the Experiment Protocol XML file between tags `<oscilloscope id="1">` and `</oscilloscope>`.

With regard to the configuration of the horizontal axis, it is common to both channels, with the capacity to configure the seconds/division (horz_samplerate parameter) and the reference position of the time axis (horz_refpos). The number of samples to be captured is a value that cannot be modified by the user and is always equal to 500 (horz_recordlength) (Code 2.5).

In the oscilloscope available in the EC, the user can perform in a virtual manner the same operations as with any manual oscilloscope found in traditional laboratories. Thus, for each of the two available measurement channels (CH1 and CH2), the user can determine whether to enable it (chan_enabled parameter) to configure the volts/division to be displayed on the screen (chan_range) and apply an offset to the display (chan_offset) if so desired. The attenuation value of the measuring cable is always 1 (chan_attenuation) (Code 2.6).

```
<horizontal>
        <horz_samplerate value="2000"/>
        <horz_refpos value="50"/>
        <horz_recordlength value="500"/>
</horizontal>
```

Code 2.5. Oscilloscope horizontal axis set-up.

```
<channels>
        <channel number="1">
                <chan_enabled value="1"/>
                <chan_coupling value="dc"/>
                <chan_range value="0.5"/>
                <chan_offset value="0"/>
                <chan_attenuation value="1"/>
        </channel>
                <channel number="2">
                <chan_enabled value="1"/>
                <chan_coupling value="dc"/>
                <chan_range value="1"/>
                <chan_offset value="0"/>
                <chan_attenuation value="1"/>
        </channel>
</channels>
```

Code 2.6. Oscilloscope vertical axis set-up.

The trigger options available on the oscilloscope front panel also resemble those of the real instrument. Thus, the student can configure which channel activates the trigger (trig_source), whether this is to be achieved via a rising or falling slope (trig_slope), the coupling (trig_coupling), the trigger threshold (trig_level), and the shooting mode (trig_mode). The options related to the timeout (trig_timeout) and delay in the trigger (trig_delay) are not modifiable by the user and always have the default value assigned (Code 2.7).

Finally, the user can perform three measurements at the same time via the two displayed channels. For each of the measurements, the protocol includes a parameter to identify the channel where the measurement is performed (meas_channel) and the type of measurement (meas_selection) (Code 2.8).

```
<trigger>
        <trig_source value="channel 1"/>
        <trig_slope value="positive"/>
        <trig_coupling value="dc"/>
        <trig_level value="0"/>
        <trig_mode value="autolevel"/>
        <trig_timeout value="1"/>
        <trig_delay value="0"/>
</trigger>
```

Code 2.7. Oscilloscope trigger set-up.

```
  <measurements>
      <measurement number="1">
             <meas_channel value="channel 1"/>
             <meas_selection value="none"/>
      </measurement>
      <measurement number="2">
             <meas_channel value="channel 1"/>
             <meas_selection value="none"/>
      </measurement>
      <measurement number="3">
             <meas_channel value="channel 1"/>
             <meas_selection value="none"/>
      </measurement>
  </measurements>
<osc_autoscale value="0"/>
```

Code 2.8. Oscilloscope measurements' set-up.

All the parameters that can be set up by the user are accessible via the buttons and knobs available at the oscilloscope from the panel. The autoscale option is disabled in the front panel (<osc_autoscale value="0"/>).

2.2.2.5 *Power supply description — request*

A description of the DC power supply set-up is included in the Experiment Protocol XML file between tags <dcpower id="1"> and </dcpower>. In the case of this instrument, the three direct voltage outputs can be configured, each of them with different voltage values (Code 2.9).

Thus, for each DC output, the desired voltage must be indicated, the limits being +20 VDC, –20 VDC, and +6 VDC. If the user tries to

```
<dcpower id="1">
	<dc_outputs>
		<dc_output channel="6V+">
			<dc_voltage value="0"/>
			<dc_current value="0.5"/>
		</dc_output>
			<dc_output channel="25V+">
			<dc_voltage value="10"/>
			<dc_current value="0.5"/>
		</dc_output>
			<dc_output channel="25V-">
			<dc_voltage value="0"/>
			<dc_current value="0.5"/>
		</dc_output>
	</dc_outputs>
</dcpower>
```

Code 2.9. Triple DC power supply set-up.

configure the outputs with voltage values higher than these, they will receive the message replicated in Figure 2.4.

On the other hand, despite the appearance in the configuration fields of the output current of each of the sources, this field cannot be modified by the user, always being 0.500 A. Nevertheless, the VISIR internally limits the maximum current to be supplied to 100 mA (0.1 A) to avoid any potentially dangerous situation.

2.2.2.6 *Multimeter description — response*

The response sent from the MS to the EC begins, as does the request frame, with identification of the protocol version and the session key ID (Code 2.10).

Following this identification, the description of the multimeter response appears. This description includes the result of the measurement

```
<protocol version="1.3">
<response sessionkey="79b9674e14a9c1005fa4c1d3cd723aa0">
```

Code 2.10. Description of Experiment Protocol version and session key at response frame.

```
<multimeter id="1">
   <dmm_function value="dc volts"/>
   <dmm_resolution value="3.5"/>
   <dmm_range value="-1.000000e+00"/>
   <dmm_result value=" 1.000237e+01 "/>
</multimeter>
```

```
<multimeter id="1">
   <dmm_function value=" resistance "/>
   <dmm_resolution value="3.5"/>
   <dmm_range value="-1.000000e+00"/>
   <dmm_result value="9.790891e+03"/>
</multimeter>
```

Code 2.11. Description of multimeter responses.

performed in the circuit under test, in accordance with the requested set-up sent previously from the EC to the MS.

As an example, Code 2.11 shows first the nature of the multimeter response in the event of having requested the reading of a DC voltage between nodes in the circuit, and the second example displays the measurement of a resistance value.

2.2.2.7 *Function generator description — response*

As we have indicated in the description of the file containing a request, the response frame always contains the same fields, regardless of whether all the instruments are used or not. The only exceptions are the fields corresponding to the circuit description. In this case, the response frame does not contain the tags <circuit> and </circuit>.

In the case of the function generator, the description contained between the tags <functiongenerator> and </functiongenerator> corresponds to the reading request including the indicated parameters (Code 2.12). *A priori*, this information should be the same as the information the user has configured on the front panel of the instrument contained

```
<functiongenerator>
    <fg_waveform value="sine"/>
    <fg_amplitude value="5.000000e-01"/>
    <fg_frequency value="1.000000e+03"/>
    <fg_offset value="0.000000e+00"/>
    <fg_startphase value="0.000000e+00"/>
    <fg_triggermode value="continuous"/>
    <fg_triggersource value="immediate"/>
    <fg_burstcount value="0"/>
    <fg_dutycycle value="5.000000e-01"/>
</functiongenerator>
```

Code 2.12. Description of function generator response.

in the EC, but in the event of any configuration error, or if the circuit causes, for example, the power signal to be saturated, this information can be very useful to the user in order to detect possible unexpected operations in the measured circuit.

2.2.2.8 *Oscilloscope description — response*

In the case of the oscilloscope, the response sent from the MS to the EC includes both the information necessary to represent the signals on the front panel display of the instrument and the result of the measurements, if the user has configured them.

As an example, the code received in the EC that makes it possible to represent the signals indicated in Figure 2.5 is shown in Codes 2.13–2.15. First, after the label <oscilloscope>, there is another indication that the autoscale option is disabled.

Code 2.13 shows the information corresponding to the horizontal axis set-up. Obtained sample is a fixed value always equal to 500 for both channels. In this example, a sample rate of 1.0e+05 corresponds to 500 μs/div. Thus, a sample rate of 5.0e+4 will correspond to 1 ms/div and so on.

Channel configuration provides all the information for representation of the signal at the client's scope. For example, the label <chan_samples> contains the information encoded in base64 of each of the 500 samples obtained during the sampling of the signals. Information about the trigger set-up is the same as that included in Code 2.7.

Finally, Code 2.15 includes the information with the reading of the measurements the user wishes to obtain with regard to the signals

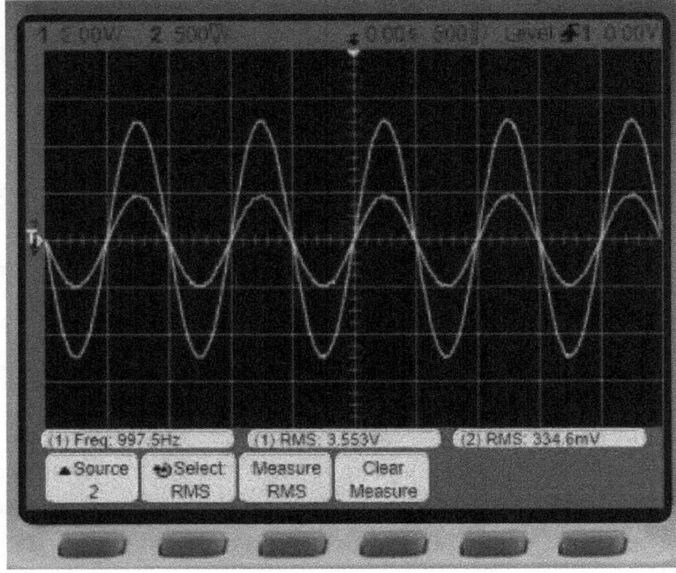

Figure 2.5. Signals displayed on the oscilloscope front panel in accordance with Codes 2.13–2.15.

```
<osc_autoscale value="0"/>
 <horizontal>
    <horz_samplerate value="1.000000e+05"/>
    <horz_refpos value="5.000000e+01"/>
    <horz_recordlength value="500"/>
</horizontal>
```

Code 2.13. Description of horizontal axis set-up at the oscilloscope response.

represented on the front panel. In the example shown in Figure 2.5, the user wants to measure the frequency and RMS voltage of the signal displayed on CH1, as well as the RMS voltage value of the signal displayed on CH2.

2.2.2.9 *Power supply description — response*

The response received at the EC with the information about the triple DC power supply is shown in Code 2.16. This code includes this instrument's

```
<channels>
   <channel number="1">
       <chan_enabled value="1"/>
       <chan_coupling value="dc"/>
       <chan_range value="1.600000e+01"/>
       <chan_offset value="0.000000e+00"/>
       <chan_attenuation value="1.000000e+00"/>
       <chan_gain value="6.601400e-02"/>
       <chan_samples encoding="base64"> Af759 (…)
       </chan_samples>
   </channel>
   <channel number="2">
        <chan_enabled value="1"/>
        <chan_coupling value="dc"/>
        <chan_range value="4.000000e+00"/>
        <chan_offset value="0.000000e+00"/>
        <chan_attenuation value="1.000000e+00"/>
        <chan_gain value="1.684300e-02"/>
        <chan_samples encoding="base64"> AP/+/(…)
        </chan_samples>
    </channel>
</channels>
```

Code 2.14. Description of channels 1 and 2's set-up at the oscilloscope response.

```
  <measurements>
       <measurement number="1">
              <meas_channel value="channel 1"/>
              <meas_selection value="frequency"/>
              <meas_result value="9.975062e+02"/>
       </measurement>
       <measurement number="2">
              <meas_channel value="channel 1"/>
              <meas_selection value="voltagerms"/>
              <meas_result value="3.552519e+00"/>
       </measurement>
       <measurement number="3">
              <meas_channel value="channel 2"/>
              <meas_selection value="voltagerms"/>
              <meas_result value="3.345540e-01"/>
       </measurement>
</measurements>
```

Code 2.15. Description of measurements' readings at the oscilloscope response.

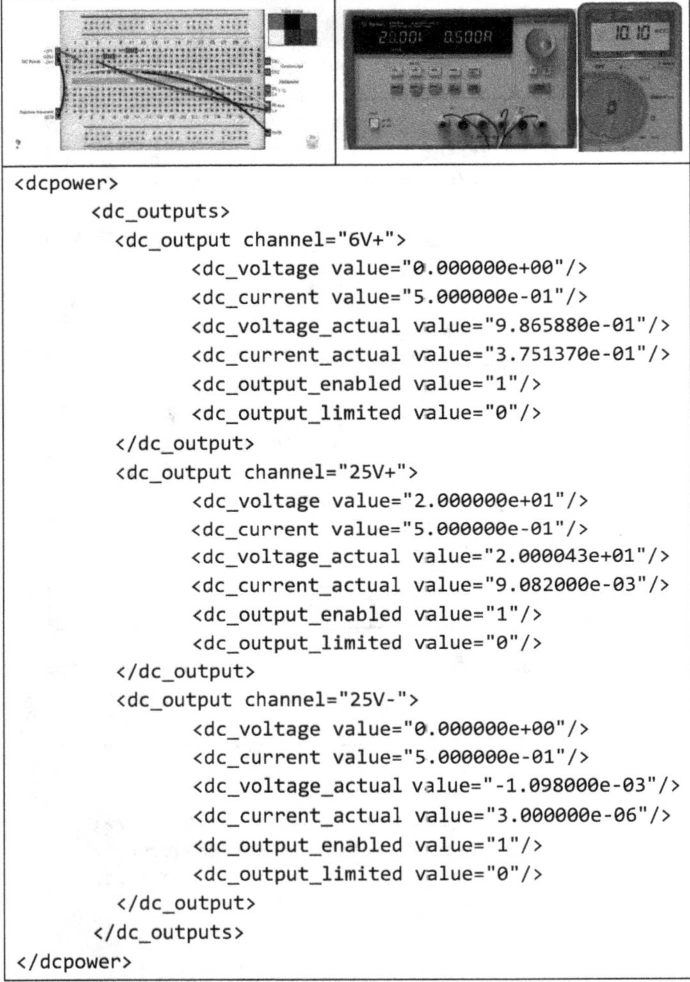

```
<dcpower>
    <dc_outputs>
      <dc_output channel="6V+">
            <dc_voltage value="0.000000e+00"/>
            <dc_current value="5.000000e-01"/>
            <dc_voltage_actual value="9.865880e-01"/>
            <dc_current_actual value="3.751370e-01"/>
            <dc_output_enabled value="1"/>
            <dc_output_limited value="0"/>
      </dc_output>
      <dc_output channel="25V+">
            <dc_voltage value="2.000000e+01"/>
            <dc_current value="5.000000e-01"/>
            <dc_voltage_actual value="2.000043e+01"/>
            <dc_current_actual value="9.082000e-03"/>
            <dc_output_enabled value="1"/>
            <dc_output_limited value="0"/>
      </dc_output>
      <dc_output channel="25V-">
            <dc_voltage value="0.000000e+00"/>
            <dc_current value="5.000000e-01"/>
            <dc_voltage_actual value="-1.098000e-03"/>
            <dc_current_actual value="3.000000e-06"/>
            <dc_output_enabled value="1"/>
            <dc_output_limited value="0"/>
      </dc_output>
    </dc_outputs>
</dcpower>
```

Code 2.16. Description of DC power supply response.

response when the user builds the circuit included in the example: The measurement of current provided to two 1 kΩ resistors connected in series with a voltage set-up of 20 V at +25 VDC source.

As can be seen, for each of the source outputs, the voltage and current values sent from the EC to the MS are included (`<dc_voltage value>` and `<dc_current value>`), as well as the reading of these values offered by the instrument (<dc_voltage_actual value> and <dc_current_actual value>).

It is important to highlight that the output current is limited to 100 mA. Therefore, if the user tries to build a circuit with a single resistance of 100 Ω and feed it with 20 V, the theoretical current that circulates through the resistor (and that is requested from the power supply) is 200 mA. However, the user will observe on the front panel of the power supply that the voltage drops to 10 V and the current measured in the circuit will be 100 mA.

2.2.3 *VISIR Measurement Server*

In view of the previous section, one could say that the main responsibility of the MS is to serve measurement requests sent by the ECs. These requests are encoded using the Experiment Protocol that contains the settings and functions of the circuit and instruments used by the student during the experiment.

However, the function of the MS (Figure 2.6) is not only to act as a gateway between the web client and the real laboratory hardware. The MS's main tasks are as follows: To manage the connections from the EC and validate the experiment requests sent from the EC; user timesharing and authentication; and communication with the EqS to execute the experiment (circuit implementation and measurement taking).

The MS is written for Microsoft Windows using Microsoft Visual C++. The dynamic properties of the XML used to encode the Experiment Protocol help add new instruments and instrument control to the system.

2.2.3.1 *Virtual instructor: <.max> files' definition*

In a traditional laboratory, the instructor is the person in charge of providing the components to the students, teaching them how to interconnect them to obtain the circuits to be analyzed in each practical session and reminding them of or introducing them to the safety rules of the laboratory to avoid damage to the equipment they are handling and to themselves. Then, during a typical hands-on session, the teacher checks each circuit to avoid possible damage. If the circuit is safe and complies with the proposed design rules, students are allowed to continue by activating the power sources, respecting the maximum values also specified by the lab

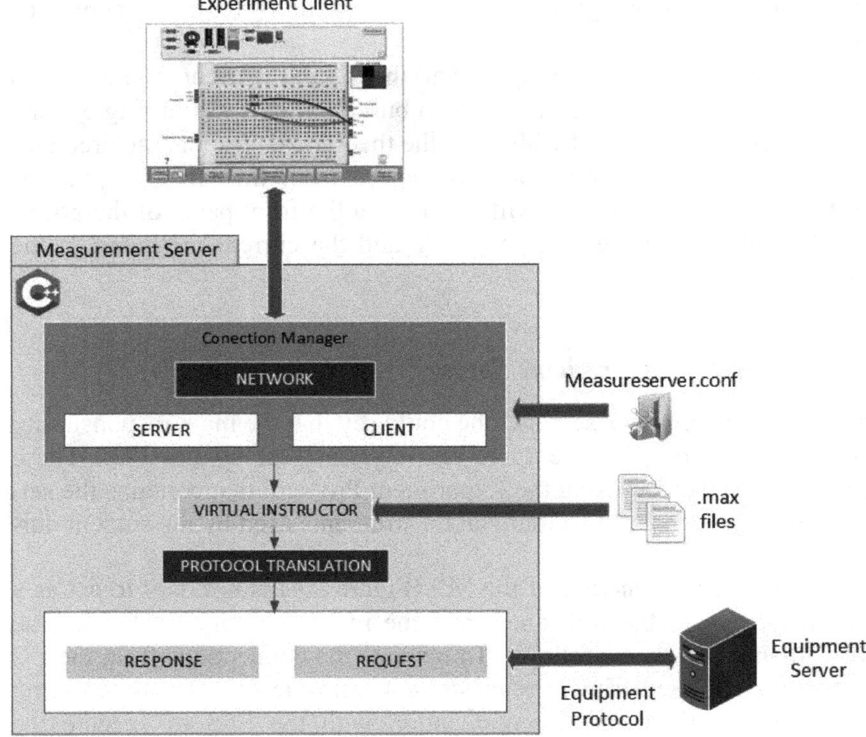

Figure 2.6. Measurement server architecture.

guide. It is true that after these first steps, once learners have acquired certain skills, it is the learners themselves who have to carry out these actions.

In summary, to avoid situations of risk, the instructor also must define those circuits that are safe and the maximum source voltages and currents that can flow through all the branches of the circuit under test without overloading any component.

In VISIR, the virtual instructor oversees these actions to avoid instrument damage. Consider what could happen in the remote lab if the user shortcuts the power supply or overloads a capacitor. It could be fatal if the capacitor burnt, even more so because the VISIR hardware is normally unattended. In a scenario like this, remote users would observe strange results in subsequent experiments involving the destroyed component. It would be worse if an instrument was damaged.

To carry out these preventative actions in VISIR, the teacher together with the laboratory technician must define the rules that determine which circuits can be built and measured by the students, as well as the maximum values of the sources used to power the circuit (AC signals through the function generator and DC signals through the triple power supply).

This set-up is achieved by defining the so-called .max files. Each of these files contains the net list that describes a circuit and the maximum values of the signals provided by the power sources. Each of these net lists is like a checklist. Thus, the checklists specify safe circuits that can be created using the components selected by the teacher and the physical wiring in the switching matrix. All subsets of a checklist included in a .max file must also represent safe circuits, but a few checklists are usually still required to maximize user freedom and permit harmless mistakes.

WARNING: To ensure optimal maintenance and scaling of the remote laboratory, it is highly recommended that each of the circuits the teacher wants to make available to the students be described in a separate .max file. VISIR makes it possible to have a single .max file with all possible net lists, but use of this option is not recommended. The different combinations of components and shortcuts that a .max file can contain may lead to uncontrolled connections that could cause dangerous situations or erroneous measurements, even if the circuit is well defined by the user on the breadboard.

Along with the checklists included in the net lists, the virtual instructor has redundant mechanisms for verifying the circuits so that it is never possible, for example, to create a shortcut by joining the positive output of a source with the ground terminal of the circuit or for the DC power supply to provide currents greater than 100 mA.

The virtual instructor routine compares each desired circuit description with all the checklists and acknowledges the circuit when it matches at least one list or a subset of a list. Moreover, the net list of components and shortcuts included in the .max files must be a sublist of components and shortcuts physically connected in the switching matrix. An example of a checklist contained in a .max file is shown in Code 2.17, which represents the circuit in Figure 2.7.

```
VDC+25V_1  F        max:20 imax:0.5
VDCCOM_1   0
VFGENA_FGENA1     A 0     max:5

D_D1         C B 1N4007  *cathode anode
D_D2         B C 1N4007  *cathode anode
R_R1         C 0 1k
R_R2         C 0 10k
C_C1          D 0 1u     *positive negative
C_C2          D 0 10u
C_C3          D 0 0.1u
SHORTCUT_S1      A B
SHORTCUT_S2      C D
SHORTCUT_S3      F A
```

Code 2.17. Example of checklist included in a .max file.

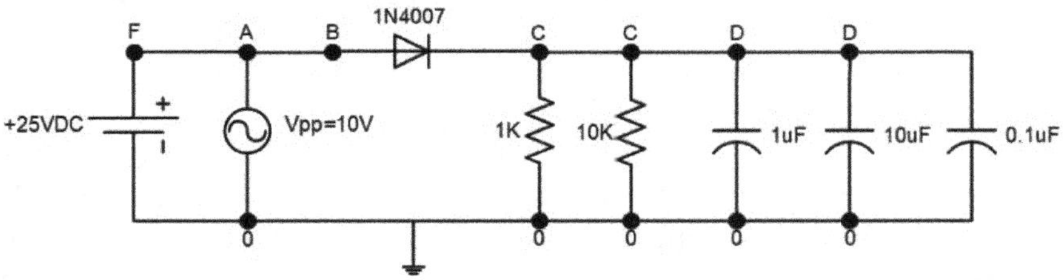

Figure 2.7. Example of circuit to create the .max file in Code 2.17.

In this .max file, the following rules are defined by the teacher in collaboration with the lab technician:

- The maximum output voltage of the function generator is ±10 V into a high impedance load.
- If the +25 VDC power is used, the maximum settings are +20 V and 0.5 A.
- If the +25 VDC output is used, the VDCCOM_1 must be connected to the ground. This means that the COM terminal at the breadboard must be connected to one of the GND connectors.
- Only the components listed in Code 2.17 can be used, and they can only be connected following the schema described in Figure 2.7. This means, for example, that the diode cannot be

connected in reverse configuration. If the teacher wants to enable this possibility, (a) another diode must be added, as underlined in Code 2.17, or (b) different shortcuts must be added to create the required branches. Detailed information about .max files and measureserver.conf files is provided in Section 2.4.1.

The role of the virtual instructor, therefore, is to determine whether the circuit that the user wants to create meets pre-established design and security requirements. In the event of non-compliance, this does not mean that the virtual instructor must show the student the errors made. Like a teacher in the classroom, you do not always have to provide the student with a detailed description of the mistake.

Thus, it is important to note that the only error message the user can receive from the VISIR is the one previously indicated (Figure 2.4). This means the error message may indicate the following:

- The user is violating one of the security rules. For example, they may be performing a shortcut that endangers the physical integrity of the lab hardware. Therefore, students must analyze the circuit to discover that failed connection.
- Some of the power equipment may be set up with a value that exceeds the maximum DC voltage or current values or amplitude of the allowed AC signals. The teacher must previously indicate these values to the students.
- If the student uses any component or connection other than those indicated by the teacher (and configured by the laboratory technician in the corresponding .max files), they will also receive this error message from the web client and will have to check the circuit to fix the possible failure.

In this way, if the user builds the circuit correctly but, for example, does not configure the power supply with the values given by the teacher, the measurements could be incorrect. In this case, VISIR does not return the error message and the user must discover the error that may have been committed.

Thus, if the user forgets to make a connection in the circuit (and the result is not dangerous for the hardware), the circuit is built with incorrect measurements, and VISIR does not return an error message, it is the user who must find the reason for the obtained results.

2.2.3.2 *User timesharing and authentication*

The timesharing scheme used to allow a certain number of simultaneous accesses with an acceptable response time imposes restrictions on the time period allowed for each experiment. The time period allowed for any single measurement is currently 100 ms. The maximum number of connections performed at the same time is a parameter that can be configured by the laboratory technician.

However, in electronics courses, the teacher can easily choose an appropriate timescale for the experiments by selecting proper values for the components to be used by the students. The time settings on the VISIR oscilloscope used in circuit analysis range from 0.5 μs/div to 5 ms/div.

The laboratory is designed for low-frequency experiments, with 1 MHz as the maximum frequency of the signals output at the function generator. Circuits assembled using the switching matrix include longer wires than those found in circuits assembled by an experienced person on a conventional breadboard. The former circuits can perhaps be compared to those built by the average student. Extra wires and relays limit the bandwidth somewhat, so the oscilloscope time base is currently restricted to 0.5 μs/div.

Authentication or user request identification is a key factor because of the limited resources. This function is understood as the follow-up to user requests when they are handled by the MS since the authentication referring to the user login is managed directly by the RLMS or by the LMS in which the VISIR access is integrated.

Therefore, as mentioned in previous sections, the request–response binomial is created through the cookie that identifies each session created by the RLMS when the user accesses VISIR.

In short, the authentication and timesharing functions are handled by the RLMS, so neither the teacher nor the lab technician needs to worry about them.

2.2.4 *Equipment Protocol*

The Equipment Protocol has been created to allow communication between the MS and the EqS and is a translation of the Experiment Protocol once the circuit and instruments' set-up defined by the user have been validated by the virtual instructor at the MS.

Like the Experiment Protocol, it is based on requests and responses. The MS builds an experiment string that defines a request and sends it to the EqS via a TCP/IP connection. The EqS then handles the request, configures the instruments, commands the switching matrix, and reads and produces a response that is sent back to the MS again in a string format. Each request is sent in a different TCP session. This means that the EqS considers each measurement configuration request as a TCP session.

Implementation of the protocol is simple and practical. It is based on two types of packets to be exchanged between the MS and the EqS: data or error packets, which have the format shown in Code 2.18, for both request and response frames.

To run an experiment, as in a hands-on lab, a certain order must be followed in the configuration and activation of the instruments. For example, we cannot take voltage measurements if we have not connected the power supply; or even more logically, we cannot take measurements if we have not built the circuit.

Therefore, the requests sent from the MS to the EqS have a specific order, in which only the instruments needed for each experiment are configured. In other words, it differs from the Experiment Protocol, where the description of all the instruments is always sent in the XML file exchanged between the EC and the MS.

This configuration is embedded in the `<content>` field indicated in Code 2.18. The first characters define the instrument addressed by the frame, following the codification and order described in Table 2.1.

This means that the first instrument to be configured is always the switching matrix and then the instruments that power the circuit, i.e., the function generator and power supply. After this configuration, a delay of

```
<Length><Type><\n><Content><\n>
   <Length> : String (6 characters long)
      The length of the data following.
      Not including length
   <Type>   : String
      Type of request: data or error
   <Content> : Data
      Packet payload
```

Code 2.18. Equipment Protocol: package format.

Table 2.1. Instrument IDs at <content> field.

<Instrument ID>	Value
Circuit Builder	41
Function Generator	11
DC Power Supply	12
Extended Peripherals	31
Digital Multimeter	22
Oscilloscope	21
Digital I/O	42
Circuit Switch	44

100 ms is introduced that serves to stabilize the signals of the circuit. It is executed via the extended peripheral instrument.

As an example, and before detailing how the configuration of each instrument is carried out, Code 2.19 shows how, from a client implemented in Java in a very simple way, it can emulate the MS sending a request for configuration of an instrument (underlined), in this case, the function generator (circled number 11 <Instrument ID>=11).

As can be seen in the example of Code 2.19, requests and responses are sent as data packets. There can be multiple instrument requests/ responses in one packet, separated by newline ('\n'). All instrument requests and responses are encoded in the style of Code 2.20, where `<Instrument Data>` field is specified in the following sections for each instrument.

The following subsections describe the information for each instrument that follows the Instrument ID description defined in Code 2.20 and integrated into the `<content>` field (Code 2.18).

2.2.4.1 *Circuit Builder*

When the Instrument ID field is equal to 41, in accordance with Table 2.1, this means that data following it are used by the EqS to command the switching matrix. Therefore, through this information, the relays of the cards integrated into the switching matrix are controlled (opened or closed). These actions create the circuit defined by users and connect the instruments they want to use for its characterization.

```
import java.io.*;
import java.net.*;
public class smtpClient {
    public static void main(String[] args) {
      Socket smtpSocket = null;
      DataOutputStream os = null;
      DataInputStream is = null;
      int first = 0;
      for (int i = 0; i < 1; i++) {
        try {
         smtpSocket = new Socket("127.0.0.1", 5001);
         os                          =                        new
DataOutputStream(smtpSocket.getOutputStream());
         is = new DataInputStream(smtpSocket.getInputStream());
        }
        catch (UnknownHostException e) {
          System.err.println(e.getMessage());
        }
        catch (IOException e) {
          System.err.println(e.getMessage());
        }
        if (smtpSocket != null && os != null && is != null) {
       try {
            os.writeBytes("000034data\n");
            os.writeBytes("11 1 2 20000 0 0 1 0 0 0,5 0\n");
            String responseLine;
            while ( (responseLine = is.readLine()) != null) {
            System.out.println("Server: " + responseLine);
            if (responseLine.indexOf("Ok") != -1) {
                      break;
                      }
                  }
          os.close();
          is.close();
          smtpSocket.close();
        }
        catch (UnknownHostException e) {
          System.err.println("Trying to connect to unknown
host: " + e);
        }
        catch (IOException e) {
          System.err.println("IOException:  " + e);
        }
      }
    }
  }
}
```

Code 2.19. Example of Java client to emulate Equipment Protocol between MS and EqS.

```
<InstrumentID><\t><Instrument Data>
   <Instrument ID> : Table 2.1
   <Instrument Data>: setup parameters
```

Code 2.20. Instrument ID description.

Table 2.2. Instruments ID on the Circuit Builder.

Functions	Value
Circuit Builder	0
Circuit Switch	1
Clear All Cards	2
Measurement Connection	3

The first field after the Instrument ID determines the function to be executed according to Table 2.2.

A Circuit Builder request represents the codification of relays that must be closed to implement via the switching matrix of the circuit built by the student at the web client. This function includes the information required to connect the components to each other, as well as the nodes to which the function generator and/or power supply is connected. The Measurement Connection function describes the circuit nodes to which the multimeter and/or oscilloscope must be connected to take measurements.

With this information and the description of the components placed in the switching matrix, the EqS composes the commands that open or close the required relays in each switching matrix board.

A Circuit Switch Request command makes it possible to toggle a specific component. Like the Clear All Card request command, these are functions that cannot be performed from the web client. These commands are available only for maintenance purposes.

As an example, Code 2.21 shows how the following circuit would first be encoded by the Experiment Protocol; then how it would be translated by the MS following the specifications of the Equipment Protocol; and finally, how it would then result in the interconnection of components and instruments on the switching matrix, in accordance with the description made by the user in the breadboard.

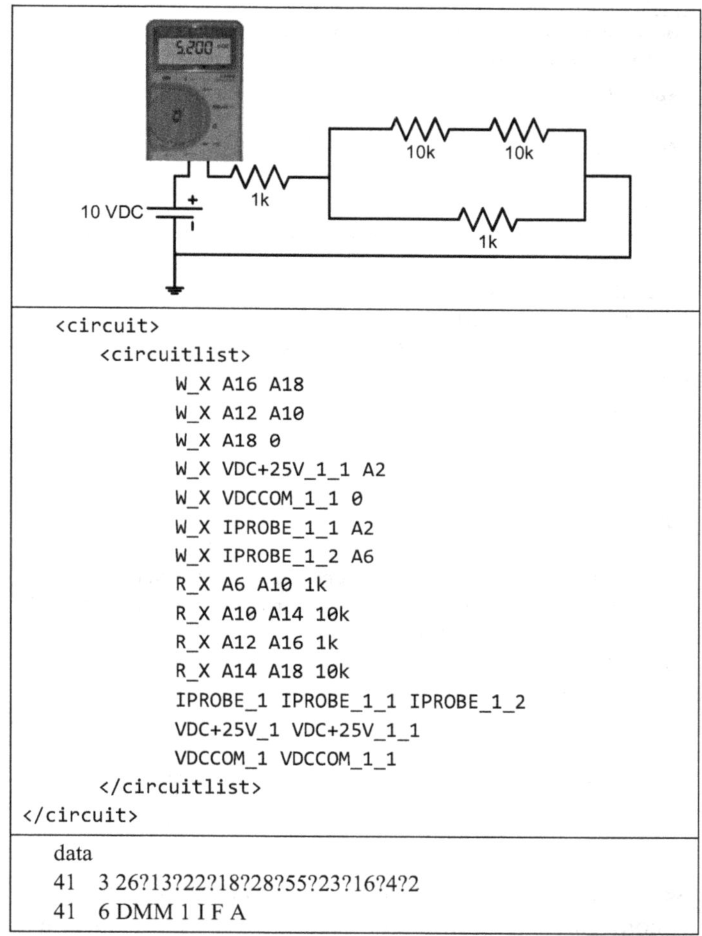


```
<circuit>
    <circuitlist>
        W_X A16 A18
        W_X A12 A10
        W_X A18 0
        W_X VDC+25V_1_1 A2
        W_X VDCCOM_1_1 0
        W_X IPROBE_1_1 A2
        W_X IPROBE_1_2 A6
        R_X A6 A10 1k
        R_X A10 A14 10k
        R_X A12 A16 1k
        R_X A14 A18 10k
        IPROBE_1 IPROBE_1_1 IPROBE_1_2
        VDC+25V_1 VDC+25V_1_1
        VDCCOM_1 VDCCOM_1_1
    </circuitlist>
</circuit>
```

```
data
41  3 26?13?22?18?28?55?23?16?4?2
41  6 DMM 1 I F A
```

Code 2.21. Example of Experiment and Equipment Protocol Circuit description.

2.2.4.2 *Function generator*

Regarding the rest of the instruments, two functions can be performed: set up (0) or fetch (1). In general, the set-up function is associated with a request action over the instrument in order to set it up, and the fetch function is associated with a response by the instruments with the reading of its internal values.

```
<Waveform><Amplitude><Frequency><DC Offset><Start phase>
<Trigger mode><Trigger source><Burst count><Duty Cycle
High><User defined waveform>

<Amplitude>  : Double
   Limits: 0 – 10.0 V peak to peak
<Frequency>  : Double
   Limits: Sine, Square = 20 MHz, All others = 1 MHz
<DC Offset>  : Double
   Limits: -5 V to +5 V, |(DC Offset + Amplitude)| < 10 V
<Start phase>: Double
   Limits: -180.0 to +180.0 degrees
<Burst count>*: Integer
   It is 0 when disabled.
   Notice: it should always be set to 0 on the 5401
   function generator. This function is only valid at 5411
   Functions generator
<Duty Cycle High>   : Double
   This property controls the duty cycle of the square wave
   the function generator is producing. You specify this
   property as a percentage of time the square wave is high
   in a cycle. Default Value: 50%
<User defined waveform>   : Array of 512 Double or '0'
   Should only be sent as a single 0, when no user waveform
   is used.
```

Code 2.22. Set-up sequence at the function generator.

The set-up function is followed by the parameters with which the user wishes to configure the instrument according to the sequence given by Code 2.22.

As stated in Code 2.22, each of the fields can have a certain range of values. The rest of the parameters are defined by a series of values as indicated in Table 2.3.

By way of illustration, Code 2.23 shows an example of a sequence received in the EqS with a request to set up the function generator.

A fetch request has the same format as indicated in Code 2.2 but without including information about `<Burst count>` and `<User defined waveform>` fields.

Table 2.3. Functions over the Circuit Builder.

Waveform	Value
Sine	0
Square	1
Triangle	2
Ramp-up	3
Ramp-down	4
DC	5
Noise	6
User-defined	7

Trigger Mode	Value
Single	0
Continuous	1
Stepped	2
Burst	3

Trigger Source	Value
Immediate	0
External	1

```
data
11 0 1 2 1350 1 0 1 0 0 0.5 0
   <Instrument ID> = 11 ; function generator
   <Function> = 0 ; setup
   <Waveform> = 1 ; Square
   <Amplitude> = 2 ; 2V so it means 4V peak to peak
   <Frequency> = 1350 ;  1350Hz
   <DC Offset> = 1 ; 1V
   <Start phase> = 0 ; 0 degrees
   <Trigger mode> = 1 ; continuous
   <Trigger source> = 0 ; immediate
   <Burst count> = 0 ; disable
   <Duty Cycle High> = 0.5 ; 50%
   <User defined waveform> = 0 ;
```

Code 2.23. Example of function generator description at Equipment Protocol.

2.2.4.3 *Power supply*

The available triple DC power supply has the following outputs (each channel has separate characteristics):

- +6 (0.0 – 6.0) V, (0.0 – 1.0) A
- +20 (0.0 – 20.0) V, (0.0 – 100 m) A
- –20 (–20.0 – 0.0) V, (0.0 – 100 m) A.

Then, each of the three outputs must be switched on (or enabled) and set up, following the sequence shown in Code 2.24.

As an example, Code 2.25 shows the sequence that requests the configuration of the two ±20 VDC outputs with their respective voltages.

```
<Enable><Voltage +6><Current Limit +6><Voltage +20>
<Current Limit +20><Voltage -20><Current Limit -20>

  <Enable>   : Integer
    0:  Disable
    1:  Enable (default)
  <Voltage>  : Double
    Valid value is described under the parameter Channel
    Name above.
  <CurrentLimit>    : Double
    Valid value is described under the parameter Channel
    Name above.
```

Code 2.24. Set-up sequence at the DC power supply.

```
data
12 0 0 0.0 0.0 1 15 0.5 0 0.0 0.0
  <Instrument ID> = 12 ; DC power supply
  <Function> = 0 ; setup
  <Enable> = 0 ; Disable +6V output
  <Voltage +6> = 0.0 ; 0 V
  <Current Limit +6> = 0.0 ; 0A
  <Enable> = 1 ; Enable +15V output
  <Voltage +20> = 15 ; +15V
  <Current Limit +20> = 0.5 ; 0.5A
  <Enable> = 1 ; Enable -15V output
  <Voltage -20> = -15 ; -15V
  <Current Limit -20> = 0.5 ; 0.5A
```

Code 2.25. Example of DC power supply description at Equipment Protocol.

The response to the fetch function sent from the EqS to the MS includes information about the voltage and current provided by each of the three power supply outputs, following the same Code 2.24 format. Note that all channels are always included in the response.

2.2.4.4 *Digital multimeter*

The digital multimeter available in VISIR permits the most common measurements that are carried out in an electronics laboratory. Thus, in the Equipment Protocol, after the definition of the requested function (set up or measure – 0; fetch – 1), the measurement to be taken, its resolution and range, and the Autozero parameter are set up (Code 2.26).

It is important to highlight that this protocol is designed to support different types and models of instruments. This means that the laboratory technician should also check whether all the set-up values are supported by the instrument model integrated into their VISIR version. Examples of this issue are the different functions supported by the Equipment Protocol, where not all are accessible through the digital multimeter front panel displayed in the VISIR web client (Table 2.4).

In the case of the multimeter, it is in the response sent from the EqS to the MS where the value read by the instrument is included, depending on the function configured in the request. As an example, Code 2.27 shows the request to measure a resistance value and the response sent in which the measured value is observed over the actual circuit.

```
<Measure><Resolution><Range><Auto zero>
   <Resolution>: valid values are
      0: 3.5 Digit precision
      1: 4.5 Digit precision
      2: 5.5 Digit precision
      3: 6.5 Digit precision
   <Range> : Double
      Maximum range of the measurement. If signal is
      outside range, the result will probably be clamped
      to range.
      Autorange = -1.0
      Example 10m = 0.01
   <Autozero>: (Integer). Available value are: -1 (Auto);
      0 (off-default) ; 1 (On) and 2 (once)
```

Code 2.26. Set-up sequence at the digital multimeter.

Table 2.4. Functions over the digital multimeter.

Measure	Value
DC volts	0
AC volts	1
DC	2
AC	3
Resistance (2 wire)	4
Resistance (4 wire)	5
Frequency	6
Period	7
Diode	8

```
Request:
data
22 1 4 0 -1 0
   <Instrument ID> = 22 ; Digital multimeter
   <Function> = 1 ; set-up
   <Measure> = 4 ; resistance measurement
   <Resolution> = 0 ; 3.5 Digit precision
   <Range> = -1 ; Autorange
   <Auto zero> = 0; off-default
```

```
Response:
data
22 0 1.003235e+3 0 0
   <Instrument ID> = 22 ; Digital multimeter
   <Function> = 0 ; fetch
   <Reading> = 1.003235e+3; value of the measurement
   <Resolution> = 0 ; 3.5 Digit precision
   <Auto zero> = 0; off-default
```

Code 2.27. Example of response and request to/from the digital multimeter.

2.2.4.5 *Oscilloscope*

As in the digital multimeter, the process of setting up and reading measurements on the oscilloscope is performed in two steps. A set-up request is sent first and then a fetch request.

The set-up request is based on Code 2.28, where each field has its own structure. As the oscilloscope has two channels, this field appears twice,

```
<Autoscale><Horizontal Conf><Channel * 2>
<Trigger><Measurement * 3>
  <Autoscale>: Boolean
  By default, this value is always 0, so Autoscale is
  disabled.
```

Code 2.28. Set-up sequence at the oscilloscope.

```
<Horizontal Conf>: <Min. Sample Rate><Reference Position>
<Record Length>
  <Horizontal Conf.Min. Sample Rate>    : Double
    Specifies the sampling rate for the acquisition.
    Units: Samples per second
    Default Value: 20 MS/s

  <Horizontal Conf.Reference Position>  : Double
    Specifies the position of the Reference Event in the
    waveform record as a percentage of the record.
    Default Value 50.0%

  <Horizontal Conf.Record Length>       : Integer
    Passes the minimum number of points needed in the
    record for each channel.
    Valid Values: Greater than 1; limited by available
    memory
    Default Value 500
```

Code 2.29. Horizontal axis set-up sequence at the oscilloscope.

while field `<Measurement>` appears three times as users can perform up to three measurements at the same time.

Set-up of the horizontal axis presents some restrictions:

(a) The combination of sampling rate and minimum record length must allow the digitizer to sample at a valid sampling rate and not require more memory than the on-board memory module permits.
(b) Regarding the reference position, when the digitizer detects a trigger, it waits for the length of time specified by the trigger delay property. The event that occurs when the delay time elapses is the Reference Event. The Reference Event is relative to the start of recording and is a percentage of recording length. For example, the value 50.0 corresponds to the center of the waveform record and 0.0 corresponds to the first element in the waveform record (Code 2.29).

```
<Channel> : <Enable>
  <Channel.Enable>  : Boolean
    If false, no more channel parameters should be sent.
    If True: <Vertical coupling><Vertical range><Vertical
    offset><Probe attenuation>
      <Channel.Vertical coupling>: Enum
      Specifies how the digitizer couples the input signal
      for the channel (AC:0; DC: 1; GND:2).
      <Channel.Vertical range>    : Double
      Specifies the absolute value of the input range for
      a channel. The units are volts.
      <Channel.Vertical offset>   : Double
      Specifies the location of the center of the range.
      <Channel.Probe attenuation>       : Double
      Specifies the probe attenuation for the input
      channel.
      Default Value 1.0
```

Code 2.30. Vertical axis set-up sequence at the oscilloscope.

Code 2.30 shows how both vertical channels of the oscilloscope are set up. If the user does not enable the channel at the oscilloscope front panel, this sequence is not sent to the EqS. Some aspects to consider during this set-up are as follows:

(a) **Vertical range**: For example, to acquire a sine wave that spans between –5 and +5 V, set the Vertical range property to 10.0 V. If the signal is outside the Vertical range, the response will be clamped to the measurement window, e.g., ±Vertical range.
(b) **Vertical offset**: The value is with respect to ground and is in volts. For example, to acquire a sine wave that spans between 0.0 and 10.0 V, set this value at 5.0 V.

Some considerations on the trigger option are as follows:

(a) **Trigger level**: The value defined for this parameter must meet the following conditions: Trigger Level ≤ Vertical range/2 + Vertical offset and Trigger Level ≥ (–Vertical range/2) + Vertical offset.
(b) **Trigger hold-off**: This affects the instrument operation only when the digitizer requires multiple acquisitions to build a complete waveform.

(c) **Trigger delay**: The trigger delay time is the length of time the digitizer waits after it receives the trigger. The event that occurs when the trigger delay elapses is the Reference Event.

(d) **Trigger mode**:
  (i) *Normal*: If no trigger is received, an error message is returned.
  (ii) *Auto*: If no trigger is received, an immediate trigger is set.
  (iii) *Auto level*: If no trigger is received, a volt max measurement is performed and the trigger level is adjusted accordingly. If no volt max level is found, an immediate trigger is set (Code 2.31).

```
<Trigger>:<Source><Slope><Coupling><Level><Holdoff><Dela
y>
<Trigger mode><Timeout>

  <Trigger.Source>: Enum
    Trigger sources are: Channel 1 (value=0); Channel 2
    (1)
    Immediate (2) or External trigger (3)
  <Trigger.Slope>: Enum
    Trigger slope can be positive (value=0) or negative
    (1)
  <Trigger.Coupling>Enum
    Trigger coupling can be AC (value=0) or DC (1)
  <Trigger.Level>: Double
    Specifies the voltage threshold for the trigger. The
    units are volts.
  <Trigger.Holdoff> : Double
    Specifies the length of time the digitizer waits after
    detecting a trigger before enabling the trigger
    subsystem to detect another trigger. The units are
    seconds.
    Default Value: 0.0 s

  <Trigger.Delay>: Double
    Specifies the trigger delay time in seconds.
    Default Value: 0.0 s
  <Trigger mode>: Enum
    Mode can be Normal (value=0), Auto (1) or Auto level
    (2)
  <Trigger.Timeout> : Double
    Defines how long the fetch function will wait for a
    trigger before it timeouts.
```

Code 2.31. Trigger set-up sequence at the oscilloscope.

Table 2.5. Available measurements at the oscilloscope.

<Measurement.Selection>	Value	<Measurement.Selection>	Value
AC estimate	1012	Positive Width	12
Area	1003	Preshoot	19
Average Frequency	1016	Rise Time	0
Average Period	1015	Rising Slew Rate	1010
Cycle Area	1004	Time Delay	1014
DC estimate	1013	Voltage Amplitude	15
Fall Time	1	Voltage Average	10
Falling Slew Rate	1011	Voltage Base	1006
FFT amplitude	1009	Voltage Base to Top	1017
FFT frequency	1008	Voltage Cycle Average	17
Frequency	2	Voltage cycle RMS	16
Integral	1005	Voltage High	8
Negative Duty Cycle	13	Voltage Low	9
Negative Width	11	Voltage Max	6
None	4000	Voltage Min	7
Overshoot	18	Voltage Peak to Peak	5
Period	3	Voltage RMS	4
Phase Delay	1018	Voltage Top	1007
Positive Duty Cycle	14		

The last field that can be set up at the oscilloscope is that of measurements. Its definition is simple, and it specifies the channel and the measurement selection following the syntaxes `<Channel><Selection>`. Measurements can be performed via the signals captured by Channel 1 or Channel 2, the permitted measurements being those specified in Table 2.5.

Regarding the response sent from the EqS to the MS to a fetch request, the format of the frame is the same as previously indicated but adding for each of the two channels the field `<Channel.Waveform>`, which contains an array of data encoded in Base64 representing the signals captured by both channels.

2.2.4.6 *Digital I/O and Circuit Switch*

Digital I/O and Circuit Switch are virtual instruments designed specifically for the VISIR, which allow the maintenance and verification actions

```
Digital I/O
   <Function> :Integer
      This function allows to clear all cards (value=1),
      clear multiple cards (3) or build a complete card (9)
         Clear multiple cards: <Card mask>
            Clear all relays multiple cards.
            <Card>    : Integer
            Which cards to clear.
         Build a complete card: <Card><Data>
            <Card>: Integer
            Which card to build a row on.
            <Data>: Integer
            Describes which relay(s) on a card to enable.
```

Code 2.32. Digital I/O configuration.

```
Circuit Switch
   <Card><Component>
      <Card>  : Integer
      Which card.
      <Component>    : Integer
      Describes which component on card to toggle (1-20).
```

Code 2.33. Circuit Switch configuration.

of the switching matrix to be carried out. They are therefore instruments that are not available to users through the EC. Digital I/O makes it possible to open all the relays on the switching matrix or only specific relays from specific component boards (Code 2.32). Circuit Switch makes it possible to open/close the specific relay that connects/disconnects a component (Code 2.33).

2.2.5 *VISIR Equipment Server*

The EqS in Figure 2.8 is the logic entity hosting the control of the instrument hardware for electronics experiments, plus the relay switching matrix. The server software is written in LabVIEW, a graphical visual programming language intended for hardware and software testing, control, and design systems, simulated or real and embedded, provided by National Instruments. The MS and the EqS can be executed on the same computer, with a logical TCP connection as the link between them.

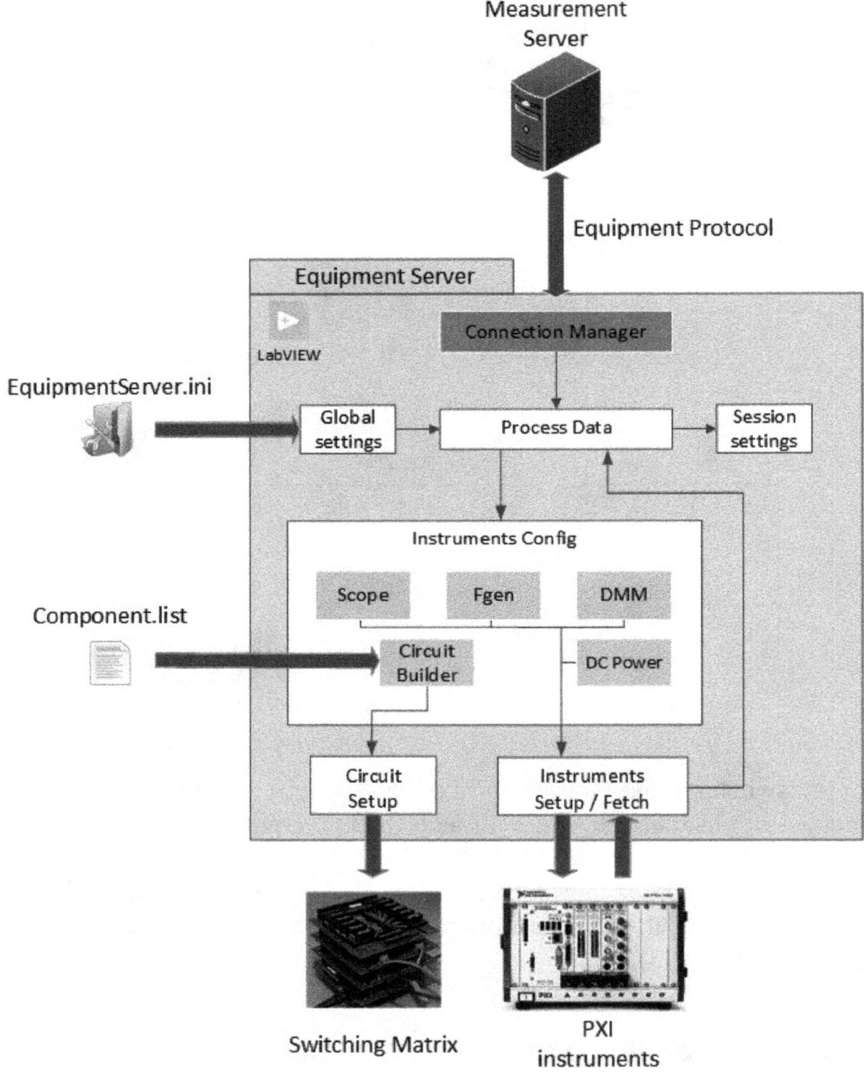

Figure 2.8. Equipment Server architecture.

The main tasks of EqS are to set up the instruments and switching matrix according to the specifications sent from the MS and described in the Equipment Protocol, and to send back the results of the measurements performed on the circuit under test.

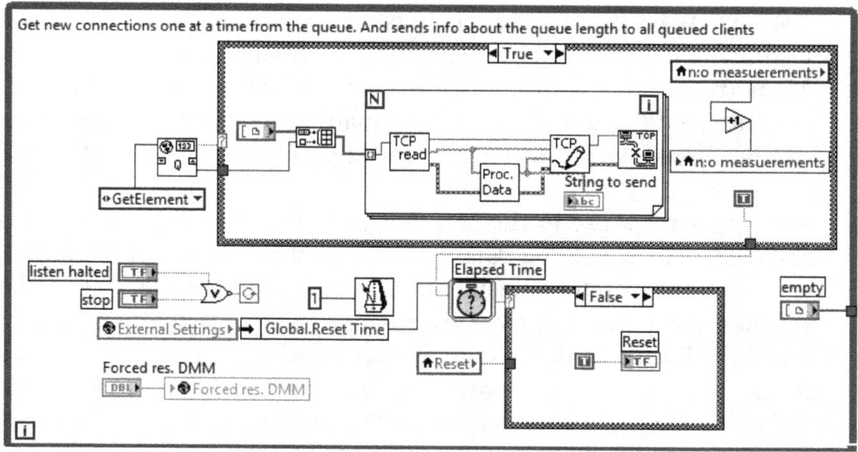

Figure 2.9. Equipment Server main loop in LabVIEW development framework.

The main functions of the main loop of EqS (Figure 2.9) can be summarized as follows:

(1) **Establish the session settings for each connection**: Starting from the configuration of each of the instruments, their state is stored in session variables so that before each configuration request by the user, the status of the instruments is checked to avoid redundancy in the configurations. This means that if a new configuration matches the actual set-up of the instrument, it is not updated, saving time in the process.
(2) **Process requests**: The data packet sent from the MS according to the Equipment Protocol format is analyzed. Depending on the information it contains, the instruments and the switching matrix are configured.
(3) **Circuit building**: Starting from the data sent by the MS in which the file made by the user in the web client is described and the `components.list` configuration file, the Circuit Builder algorithm obtains the necessary instructions to control the switching matrix and obtain the physical implementation of requested circuit.
(4) **Instrument configuration**: The set-up of the instruments is performed after checking the status of each element. The switching matrix makes the necessary connections between components and

instruments based on the circuit description sent from the MS. Each instrument is controlled by a specific algorithm.

(5) **Measurements' readings**: Once the physical circuit has been implemented through the switching matrix, measurements are taken on the circuit under test using the digital multimeter and/or the oscilloscope.

(6) **Sending a response to the MS**: Each of the instruments generates a response, which may be data or an error. When there are no errors, each instrument returns results in different ways. In the case of the function generator and the power supply, the configuration state is returned (fetch). In the case of the oscilloscope and multimeter, the reading of the measurements taken is returned. In the case of the switching matrix, a code indicating that everything has gone according to plan is returned. If an error occurs in the configuration of any of the instruments, an error packet will be returned to the MS, which contains the error code generated by the instrument and the description of that error.

Equipment Server needs two configuration files. `EquipmentServer.ini` file is described in detail in Section 2.4.1 as it is part of the lab set-up, but `components.list` file requires special attention as it is specially defined for the VISIR switching matrix. This file is described in detail in Section 2.4.2.1.

2.3 VISIR Hardware Description

The main hardware elements of the VISIR remote lab are the instruments placed in the PXI chassis provided by National Instruments and the switching matrix designed for the VISIR.

In the case of the VISIR instance installed at the University of Deusto, the available equipment integrated into the PXI chassis accessible from the EC front panels comprises the following:

- **A dual channel oscilloscope**: Whose virtual front panel represents an Agilent 54600B and most of its functions, except for "run mode" and autoscale, are supported. This front panel controls a 100 MS/s 8-bit oscilloscope NI PXI-5112 integrated into the PXI chassis.

- **A digital multimeter**: Only a preliminary virtual front panel of a multimeter is currently available. Floating two-terminal measurements are supported. The plug-in board is the 1.8 MS/s, 6½-digit multimeter NI PXI-4070, allowing for AC/DC voltage and current measurements in addition to resistance readings.
- **A function generator**: Whose virtual front panel mimics Agilent 33120A. Sine, square, triangular, and ramp functions are currently supported. The plug-in board is the 20 MHz function generator NI PXI-5412.
- **A triple DC power supply**: Whose front panel represents an Agilent 3136A. Commands sent from the EC control the PXI-4110, a 3-Channel, 20 V, 1 A PXI Programmable Power Supply.

This is just one example of instrument models that can be used but can also be replaced by others. For instance, in the VISIR instance installed at the Universidad Nacional a Distancia de Costa Rica, the oscilloscope model installed is the NI PXI-5402, the function generator model is the NI PXI-5114, and two multimeters are available: an NI PXI-4070 and an NI PXI-4072.

Along with this hardware, it is necessary to have a computer, which will exercise the functions of a server, hosting the logical entities of the MS and the EqS. This computer does not require special characteristics apart from running a LabVIEW license, as the EqS needs this software to be executed.

To facilitate the maintenance and updating tasks that may be necessary, it is recommended that a conventional computer be used and not an embedded system such as the one employed in the version installed at the University of Deusto, in which the MS and EqS are executed on the NI PXI-8105 embedded controller, as seen in Figure 2.10.

2.3.1 *Breadboard vs. Switching Matrix*

An electrical circuit consists of several connected components and at least one source. Figure 2.11 shows a graph that represents a circuit with five nodes and 10 branches. Each branch can be used to represent a component with two leads. If the number of nodes increases, so does the complexity of the electrical circuit. Asumadu *et al.* (2002) presented a small virtual breadboard and relay switching matrix combination with four nodes. If a desired circuit has N nodes and if the experimenter wishes to add, for

Figure 2.10. VISIR at the University of Deusto.

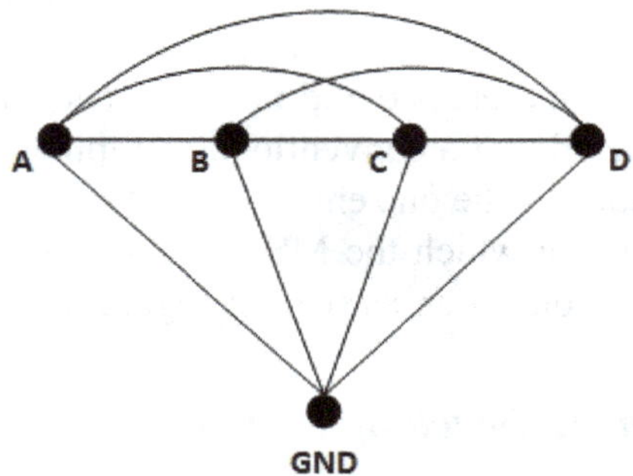

Figure 2.11. Example of a circuit with five nodes and 10 branches.

example, one resistor, there are $N \cdot (N - 1)/2$ branches into which it can be installed. This means there are 120 possibilities for 16 nodes.

However, the number of circuits that can be wired on VISIR's breadboard is limited primarily by the number of components the technician provides in the `circuit.list` file and is usually too high for a switching matrix of practical size (Scapolla *et al.*, 2005). However, in undergraduate lab exercises, students are expected to wire only simple circuits that are described in lab instruction manuals. Thus, thanks to its nine nodes (A to I) and ground node, VISIR makes it possible to create circuits with a great level of complexity (45 branches).

It should also be borne in mind that in VISIR, the laboratory technician does not have to enable the creation of all possible branches during the process of setting up the circuits (through the `.max` files).

If the user tries to make an unpermitted connection (not available in VISIR), the VISIR client shows an error message. This message will be the same if the user tries to make a connection that places the integrity of the laboratory at risk, such as shortcutting the output of the power supply. These situations are permitted because inexperienced students should be allowed to make wiring mistakes, for example, and have the opportunity to learn how to correct them. Apart from the circuits in the instruction manuals, a switching matrix should allow the student to create similar circuits that are also safe.

2.3.2 *VISIR Switching Matrix*

The switching matrix is a stack of PC/104-sized boards as shown in Figure 2.12. PC/104 is a common international standard for embedded

Figure 2.12. VISIR switching matrix: Source board, function generator and scope board, multimeter board, and components board (from bottom to top).

systems (https://pc104.org/). The relays are arranged in a three-dimensional matrix pattern together with instrument connectors and component sockets. In this way, the relay switches are embedded into the circuit created, limiting the length of the wires to gain bandwidth.

The nodes of the switching matrix are propagated from board to board, creating a node bus. The notation "node" refers to the fact that every conductor created by these stacked connectors can be a node in a desired circuit. The nodes are divided into two groups: One contains the nodes denoted as A–I and 0. The other contains nodes denoted as X1–X6 and COM (Figure 2.13). A total of 17 nodes. The nodes of the first group can be connected to the multimeter or the oscilloscope using one switch only. This means these two instruments can be connected at any of the A–I and zero nodes dynamically or by default. However, power supply and function generator can only be connected to the nodes defined in the `components.list` file as will be explained in the configuration section. X1 node is reserved to connect the +6 VDC output of the power supply, X2 to connect the +25 VDC output, and X3 to connect the –25 VDC output,

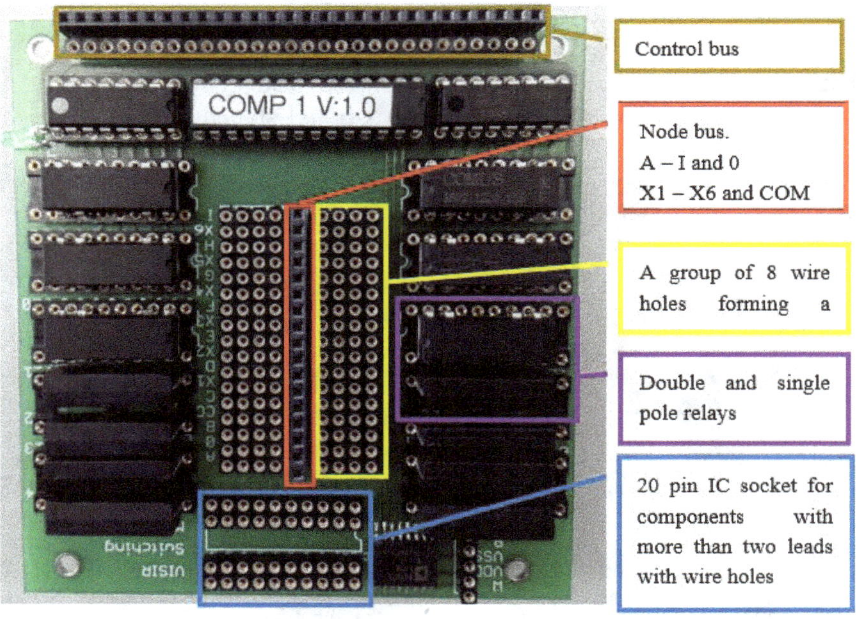

Figure 2.13. Components board version 4.1E.

while X4 is reserved for an auxiliary power supply, if needed, and X5 plus X6 are not used.

Boards also contain a control bus used to address the commands that are processed in each board to open and close relays according to the circuit to be created.

The switching matrix should be positioned onto the instrument chassis to keep the instrument cables short (Figure 2.10). From a maintenance perspective, putting the switching matrix into a closed case is not recommended because it should be easy to swap components and rewire branches.

Three types of boards have been defined so far: One for connecting sources (power supply and function generator), one for connecting oscilloscope and multimeter (Figure 2.14), and two for carrying components (Figure 2.13). A switching matrix usually contains one source board, two instrument boards, and several component boards.

The source board (Figure 2.14, left) is also the interface between the matrix and the EqS through a USB connection. A microcontroller installed in this board processes and addresses the commands to the corresponding board. These commands indicate the relays that must be closed or opened in each board, according to the description of the circuit sent from the client and validated previously by the MS. All this set-up is completely transparent to the user and technician of the VISIR remote lab. Two types of microprocessors control the switching matrix: one matrix controller and one board controller on each board. The switching matrix controller hosted on the source board connects to the board controllers via an I2C bus.

Figure 2.14. From left to right: Source board, multimeter, and oscilloscope board.

Multimeter and oscilloscope boards have the same footprint, except for the connectors to the instruments, as each instrument has its own connector (screw terminals for a multimeter and MCX connectors for connecting an oscilloscope). It is also important to consider that on the oscilloscope board, both channels (A and B) have the ground connected to a common ground by hardware. This is also the reason why at the client interface, there is no ground connection at the oscilloscope channels.

The VISIR hardware architecture deployed at the University of Deusto, with eight component boards on the top, is depicted in Figure 2.10. The bottom board is the source board. It has inputs for a triple power supply, e.g., NI PXI-4110, and a function generator, e.g., NI PXI-5402. The instrument boards can be configured for connecting a low-frequency instrument, such as a multimeter, e.g., NI PXI-4070, or for connecting a high-frequency one, such as a dual channel oscilloscope, e.g., NI PXI-5112. If two multimeters are to be installed to measure voltage and current simultaneously, it will be necessary to have two such cards in the switching matrix.

The component board version 4.1E (Figure 2.13) carries sockets where components can be installed. One component board can hold ten components with two leads, or as many components with more than two leads as can be installed in the on-board 20-pin Integrated Circuit (IC) socket. The more component boards are stacked in a switching matrix, the more components can be online. All the components connected to these boards, and their connections to the nodes, must be described in the `components.list` file.

The component board version 4.1B already shown in Figure 1.18 arranges only two components with two leads. This board is built to allow these two components to be connected to more than two nodes without needing to wire each component to the corresponding nodes. This will be explained in detail in the VISIR configuration section.

The switching matrix makes it possible to perform the circuit creation much faster than on a physical breadboard. Access to the matrix is based on a time-shared schema. Each time slice is set to 100 ms. First, the desired circuit is created and then a delay of 25 ms is allowed for the switching transients to disappear before any measurements are made. The operate/release time for the relay switches is less than 2 ms. Thus, approximately 70 ms can be used for measurements. The short time slice is a trade-off, making it possible to support many users simultaneously. The relays have reed switches. According to the datasheet, the maximum carry

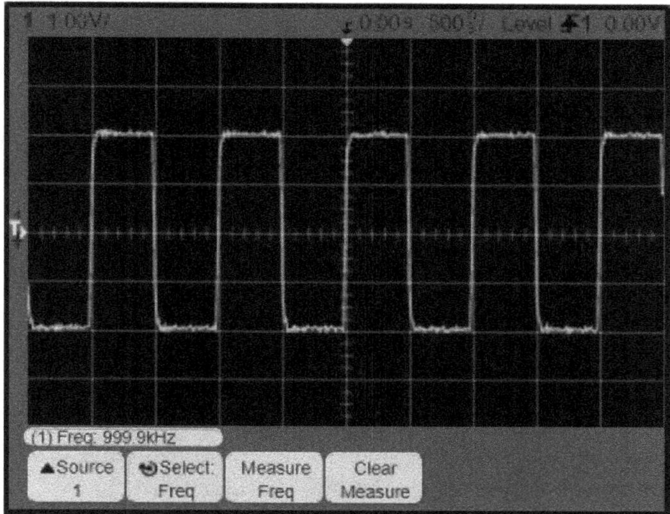

Figure 2.15. Switching matrix bandwidth test.

current for them is 2 A and the minimum life expectancy is $3 \cdot 10^8$ operations, meaning approximately two operations per second continuously for 5 years.

The switching matrix is designed for low-frequency experiments. The layout of the board strips, the wiring on the component boards, and the number of boards in the stack limit the bandwidth of the switching matrix. The result of a test of the bandwidth of the switching matrix in the workbench at the University of Deusto equipped with eight boards is shown in Figure 2.10. The function generator, NI PXI-5412, is connected to the oscilloscope, NI PXI-5112, using this matrix. A 1-MHz square wave signal is displayed on the oscilloscope. Figure 2.15 is a screen dump from a client PC.

2.4 VISIR Configuration

This section details the configuration steps required to operate the VISIR remote laboratory. The boot sequence requires that the EqS is always started first and then the MS. This order must be followed because when the MS is started, it first checks whether the EqS is operational and thus allow users to begin a session.

The MS then proceeds to load the .max files defined by the lab technician and request the EqS the `components.list` to verify that the components and instruments described in the .max files are defined in that file. If so, users can start to perform experiments. If not, an error is reported at the MS console, VISIR is not started, and the lab technician must then ascertain what is wrong with the .max files.

2.4.1 *Measurement Server Set-up*

The MS uses the following configuration files:

- **measureserver.conf**: The main MS set-up file.
- **maxlists.conf**: This file specifies the set of .max files that will be used by the virtual instructor to determine whether or not the circuit sent by the user can be built on the matrix.
- **.max files**: A set of files that describes the circuit defined by the teacher to be performed by the students.

First, `Port` and `HTTPPort` commands indicate which ports the MS is listening to connection requests from the EC. Therefore, in the EC configuration, these values must match.

Commands `MaxClients`, `MaxSessions`, `SessionTimeout`, and `Timeout` define the maximum number of clients and sessions VISIR can manage at the same time. Both timeout commands define the maximum number of seconds the MS waits for requests from the EC to kill the session (`SessionTimeout`) and the maximum time the MS waits for EqS to obtain a response from the instruments to stop the session. If this time expires, an error message is sent to the EC. The lab technician must also indicate in this file the directory to save the log files.

Finally, communication with the EqS must be set up. In the example of Code 2.34, the EqS is running in the same machine (localhost IP address is defined) and it listens to MS requests at port 5001.

Code 2.35 shows an example of the `maxlists.conf` file, which shows the files located in maxlists directory that must be loaded by the MS. These are the files used by the virtual instructor during the lab session. This means that the teacher and the lab technician can define a large set of .max files but only use a subset of these during the lab session. It reproduces the same action as in the hands-on lab when in each practical class, students only need to work on the circuits provided by the teacher in the lab session script.

```
# Comments begin line with: #
### Local configuration
# If no port configuration is given, the service will not
start
Port                   2324
HTTPPort               18080

MaxClients 10000
MaxSessions 50000
SessionTimeout 150
Timeout 100

# Max list configuration file, should contain a list of
maxlists to load
MaxListConfig          maxlists.conf

# Enable logging
Log                    1

# Logs directory must exist for the logging to work
LogDir        logs

# 1-5, 5 being the most verbose
LogLevel     5
### Equipment server module configuration
UseEQ  1
EQ.Host                        127.0.0.1
EQ.Port                        5001
```

Code 2.34. measureserver.conf file example.

```
maxlists/BJT.max
maxlists/zener.max
maxlists/AO_diferencial.max
maxlists/AO_low_pass.max
maxlists/Diodes.max
```

Code 2.35. maxlists.conf file example.

2.4.1.1. *.max files' set-up*

Finally, the lab technician, in collaboration with the teacher, needs to create all the .max files that describe all the circuits that can be created in VISIR using the components included in the `components.list` file in the EqS.

```
VFGENA_FGENA1A 0     max:5
VDC+25V_1      F     max:15 imax:0.5
VDCCOM_1       0

Q_Q1 H C B BC547C     *collector || base  || emitter

C_C1  A C 10u
C_C2  B 0 10u
C_C3  H E 10u

R_R1  F C 5.6k
R_R2  C D 1k
R_R3  F H 820
R_R4  B 0 220
R_R5  G 0 100

SHORTCUT_S1 D 0
SHORTCUT_S2 B G
```

Code 2.36. BJT.max file example.

By way of example, Code 2.36 describes the circuit shown in Figure 2.16. In this circuit, a three-lead component is used, e.g., a bipolar transistor, where its collector is connected to node H, base to node C, and emitter to node B. This sequence is quite important. Two shortcuts are used: one between nodes D and 0 (ground/common) and the second between nodes B and G.

The use of shortcuts needs to be clear from the point of view of operation, maintenance, and scalability of the system. Issues to consider are as follows:

- To connect a shortcut to the circuit, it is only necessary to activate a single pole relay. Therefore, they occupy "little" space.
- There are "special" nodes such as node A, F, or 0, that is, nodes where the power supplies and the common ground are connected. As a tip, it is better not to connect components directly to these nodes and use shortcuts for this purpose. In the example in Figure 2.16, R2 is grounded via the shortcut between nodes D and 0. This will allow us to reuse this 1k resistor in other circuits where we do not need it to be grounded.

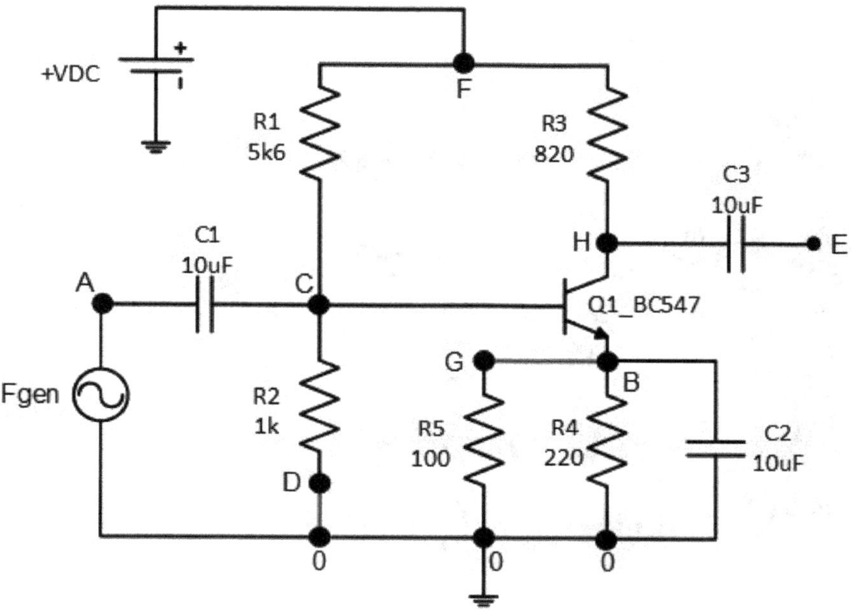

Figure 2.16. Bipolar transistor circuit.

- As a tip, having a good list of shortcuts will allow us to reuse components between some circuits and others. Thus, for example, resistance R3 of 820 Ω can be used in any other circuit where we need it to be connected between nodes F and H.
- Shortcuts allow us to replace some components with others in a simple way. In the example, we can use as emitter resistance either R5 or R4, thanks to the shortcut between G and B.
- Important: For measuring currents, as in the hands-on laboratory, the multimeter must be placed in series on the branch in which you want to take this measurement. Therefore, in VISIR, current measurement can only be taken in those places where we can open the circuit and insert the multimeter, and this is only possible when there is a shortcut. Therefore, to measure current, we will have to introduce a shortcut in that branch where we want to allow this measurement. If there is no shortcut, it is not possible to measure current. Thus, in the example, only the current circulating through the resistance R2 could be measured, placing the multimeter between this resistance and ground.

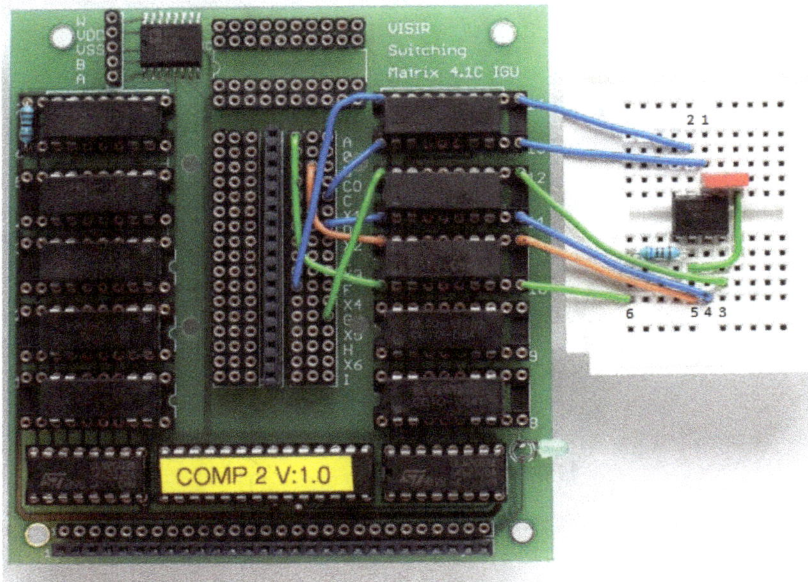

Figure 2.17. Predefined circuit based on operational amplifier.

In VISIR, it is also possible to use pre-built circuits, that is, circuits or parts of circuits that cannot be modified by the user, or where the teacher only allows the students to modify a component, for example, the feedback resistor into an inverter amplifier with an operational amplifier (Figure 2.17).

In this example, the predefined circuit is considered to be a six-lead component and can be included in a .max file as a new type of component (i.e., INVERTER) as follows: INVERTER_IV1 C F G D B A.

Finally, VISIR also provides dual component boards (Figure 1.18). In these types of boards, each of the two components that can be placed in it is connected between the following nodes: A–0, A–D, A–F, A–H, D–0, D–F, D–H, F–0, F–H, and H–0. If, for example, a 10-kΩ resistor is connected to this board, this means that in the .max files, we can include a 10-kΩ resistor connected to one of these 10 pairs of nodes.

2.4.2 *Equipment Server Configuration*

As can be seen in Figure 2.8, EqS requires two configuration files. The first, the `EquipmentServer.ini`, is used to define the global settings of

```
[Global]
Port=5001
Reset Time = 300 [s]
Maximum Delay = 1000 [ms]

[Log]
Log Level = 1
Log File = C:\VISIR\equipmentserverHTML5\EqS_log

[Instrument Address]
Function Generator 1 = PXI1Slot7
DC Power Supply 1 = PXI1Slot5
Digital Oscilloscope 1 = DAQ::4
USB   Interface   1   =   USB0::0x1043::0x0000::NI-VISA-
10002::RAW
Digital Multimeter 1 = PXI1Slot2

[Digital Multimeter]
Digital Multimeter Card 1 = 17
Digital Multimeter Card 2 = 18

[Component type]
Component type =
R:2,C:2,D:2,VDCCOM:1,OP:8,VFGENA:2,SHORTCUT:2,VDC25V:1,V
DC-25V:1,VDC6V:1,L:2,Q:3,IC:16,IC:8;INVERTER:6

[Components list]
Components list = components.list

[Matrix version]
Matrix version = 4.1
```

Code 2.37. EquipmentServer.ini configuration file.

the lab. Code 2.37 describes the configuration file of the VISIR deployed at the University of Deusto.

This file is divided into different sections denoted by square brackets:

- **Global section:** In this section, the laboratory technician must define the port used in the TCP connection between MS and EqS. This port must be the same as is indicated in the `measureserver.conf` file (Section 2.4.1). `Reset Time` and `Maximun Delay` are parameters used by the EqS to reset the connection with the MS in the event of inactivity or if a response is lost to avoid any undefined state.

- **Log section:** This section is used to track information about the connection between the MS and EqS. Log level can be set from one (less information) to five (more information).
- **Instrument address:** This section provides information to EqS about which slot of the PXI each instrument of the VISIR is connected to. To facilitate this task, use of the Measurement & Automation Explorer application provided by National Instruments is recommended. This software displays all the hardware connected to the server. Figure 2.18 shows the hardware configuration of VISIR at the University of Deusto, which corresponds to the definition included in Code 2.8. It is important to highlight that the instrument named "`USB Interface 1`" corresponds to the switching matrix.
- The instrument card number for the multimeters is given in the following. In the case of the example of Code 2.37, it is indicated

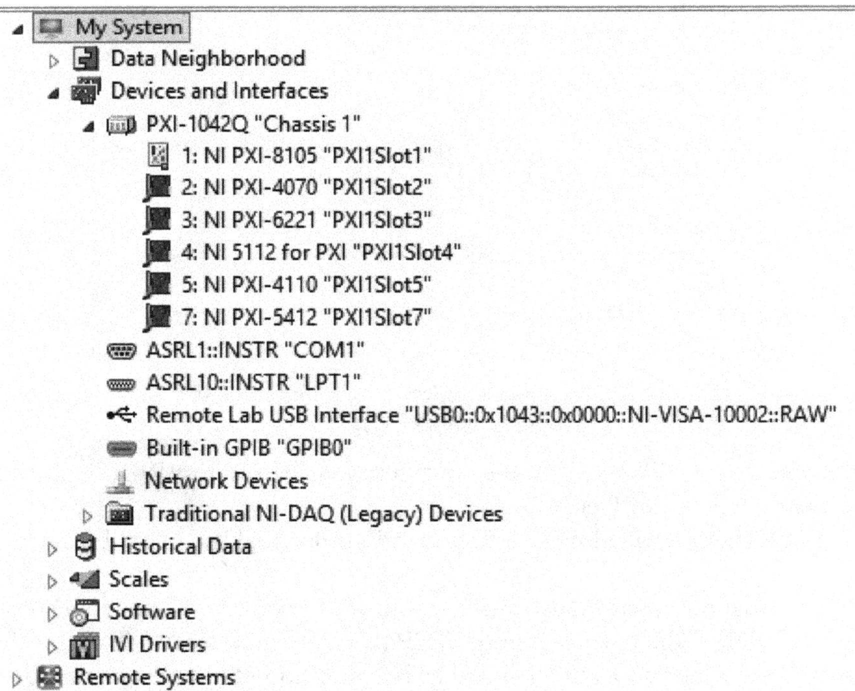

Figure 2.18. Hardware configuration of VISIR at the University of Deusto — Measurement & Automation Explorer screenshot.

there is a multimeter on card 17 (as seen in Figure 2.14) and there would be another multimeter that would be connected to instrument card 18.

- The next section introduces the types of components that are available in the matrix and that are included in the file `components.list`.
- Finally, the name of the file that contains the description of the components connected in the matrix is defined (`components.list`) as is the matrix version.

2.4.2.1 *components.list file description*

The most important element of the VISIR is its switching matrix, which is explained in detail in Section 2.3.2. It is in the matrix that the laboratory technician places the components that the user can choose in the EC to build the experiments. However, in addition to making those connections in the matrix, the technician must configure the `components.list` file. An example of this file is shown in Code 2.38.

This file aims to identify the relays the Circuit Builder must command to connect the circuit to the instruments or the components selected by the user. As can be seen, the multimeter and the oscilloscope do not appear because these instruments can be connected to all circuit nodes, so their connection to one node or another will depend on the position selected by the user for taking measurements.

```
*INSTRUMENTS SETUP
VFGENA_24_1           A 0
VDC-25V_24_5:1_11     G
VDC+25V_24_4:7_14     F
VDC+6V_24_3:3_11      E
VDCCOM_24_2           0

*COMPONENTS SETUP
R_11_13   A B 1k       *R1
D_11_3    C B 1N4007   *cathode anode
C_11_10   D 0 1u       *positive negative

SHORTCUT_11_7    F C      *S1
```

Code 2.38. components.list file example.

In the case of the function generator and the power supply, the laboratory technician must indicate to which nodes of the circuit they will be connected. Thus, in the example, it is indicated that the output of the function generator will only be connected to node A. That is, if we want to design a circuit using the function generator as an AC power supply, some of the components must be connected to node A, directly or using a shortcut.

The relays used to connect the function generator and the DC power source are defined by the hardware design of the switching matrix. However, the lab technician must define for the rest of the available components the board and the relay to which the component is connected. By way of example, Figure 2.19 represents the components described in Code 2.34. In this example, a 1 kΩ resistor is connected between nodes A

Figure 2.19. Example of components' connection following Code 2.34.

and B using the double pole relay number 13. A shortcut is created between nodes F and A when single pole relay number 7 is closed. Both components are placed on component board number 11.

The type of components (shortcuts are also considered components) that can be used in the matrix must be predefined in the EquipmentServer.ini file, as explained in Section 2.4.1.

It is important to remember that .max files must contain a subset of the components included in the `components.list` file. If this is not the case, when the MS is initialized by the lab technician, an error message will be displayed.

As shown in Figure 2.19, the laboratory technician must make the physical connections defined in the `components.list` file. That is, if it is stated that the 1 μF capacitor is connected to nodes D and 0 through relay 10 of component board 11, the technician must connect that relay to the nodes indicated, as has been done in the example with the red and green wires.

In case of a dual-component board, connections between nodes and the two components are already implemented by a set of relays. Thus, the laboratory technician does not need to wire the components with the nodes, as has been explained.

An example of the code that should be added in the `components.list` file, to describe to which nodes a 6.8 mH inductor and a 22 nF capacitor can be connected, is included in Code 2.39.

Finally, if one wishes to introduce a predefined circuit such as the one shown in Figure 2.17, it must be considered to be a component, in this example, of six leads. Then, for its connection to the circuit, three double-pole relays can be used, as described in the `components.list` file, in the last line of Code 2.39, meaning that this inverter is on component board number 2, and relays 13, 11, and 10 will be closed to connect to the correspondent nodes.

2.4.3 *Experiment Client Set-up*

The EC configuration file is named `library.xml`. In this file, the laboratory technician needs to set up the list of components to which the user has access at the web client. An example of this set-up is shown in Code 2.40.

```
* Dual component board 13
L_13_18:13_1        A 0 6.8m_X
L_13_18:13_3        A D 6.8m_X
L_13_18:13_5        A F 6.8m_X
L_13_18:13_7        A H 6.8m_X
L_13_4:13_1         D 0 6.8m_X
L_13_4:13_5         D F 6.8m_X
L_13_4:13_7         D H 6.8m_X
L_13_6:13_1         F 0 6.8m_X
L_13_6:13_7         F H 6.8m_X
L_13_8:13_1         H 0 6.8m_X

C_13_17:13_16       A 0 22n_X
C_13_17:13_14       A D 22n_X
C_13_17:13_12       A F 22n_X
C_13_17:13_10       A H 22n_X
C_13_13:13_16       D 0 22n_X
C_13_13:13_12       D F 22n_X
C_13_13:13_10       D H 22n_X
C_13_11:13_16       F 0 22n_X
C_13_11:13_10       F H 22n_X
C_13_9:13_16        H 0 22n_X

INVERTER_2_13:2_11:2_10 inverter
```

Code 2.39. Code describing dual components in the components.list file.

In this `library.xml` file, each component description contains the following:

- Type, value, and pins: The component type must be one of those defined at the EquipmentServer.ini (Code 2.37). Value corresponds to the value of the component included in the `components.list` file and pins value determines the number of leads in this component, also defined in the EquipmentServer.ini.
- Each component can be rotated in the interface or not. Rotation is an attribute that can be defined by the laboratory technician during component set-up. To do this, under the `<rotations>` tab, the position that the component takes must be defined. Values of "`ox`" and "`oy`" are defined as the center point of the image that

represents the component, measured in pixels and considering the top right corner of the image as reference position (0, 0). Thus, the center of the image reference now being (0, 0), positions of component pins are defined (Figure 2.20).

```
<component type="R" value="220k" pins="2">
<rotations>
   <rotation ox="-27" oy ="-7" image="r_220k.png" rot="0">
       <pins><pin x="-27" y="0" /><pin x="27"  y="0" /></pins>
   </rotation>
   <rotation ox="-7" oy ="-27" image="r_220k.png" rot="90">
       <pins><pin x="0" y="-26" /><pin x="0" y="26" /></pins>
   </rotation>
   </rotations>
</component>
<component type="D" value="1N4002" pins="2">
<rotations>
   <rotation ox="-27" oy ="-7" image="d_1n4002.png" rot="0">
       <pins><pin x="-26" y="0" /><pin x="26"  y="0" /></pins>
   </rotation>
   <rotation ox="-7" oy ="-27" image="d_1n4002.png" rot="90">
       <pins><pin x="0" y="-26" /><pin x="0" y="26" /></pins>
   </rotation>
   <rotation ox="-27" oy ="-7" image="d_1n4002.png" rot="180">
       <pins><pin x="26" y="0" /><pin x="-26"  y="0" /></pins>
   </rotation>
   <rotation ox="-7" oy ="-27" image="d_1n4002.png" rot="270">
       <pins><pin x="0" y="26" /><pin x="0" y="-26" /></pins>
   </rotation>
</rotations>
</component>

<component type="C" value="0.1u" pins="2">
<rotations>
   <rotation ox="-19" oy ="-24" image="c_0.1u.png" rot="0">
       <pins><pin x="-13" y="13" /><pin x="13"  y="13" /></pins>
   </rotation>
</rotations>
</component>
```

Code 2.40. library.xml file describing available components at the experiment client.

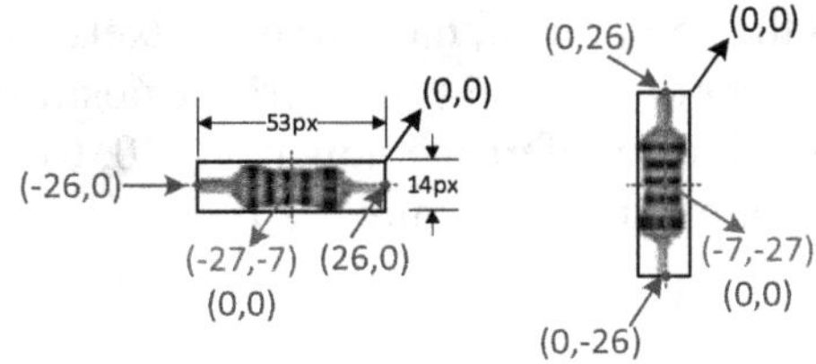

Figure 2.20. Example of resistance position calculation (rot = 0° and rot = 90°).

All the components included in the `components.list` file must be included in the `library.xml`. If they are not in this file, they are not accessible for the user at the EC.

This is the only file that must be set up at the EC. As the VISIR remote lab is embedded into the WebLab-Deusto RLMS, it is this management system that manages the connections between the EC and the MS.

References

Asumadu, J. A., Tanner, R., Fitzmaurice, J., Kelly, M., Ogunleye, H., Belter, J., & Chin, S. (2002). Nuts and volts: A web based hands on electrical and electronics remote wiring and measurement laboratory (RwmLAB). In *Proceedings of the 2002 ASEE Annual Conference*, Montreal, Canada, June 16–19.

Orduña, P., Gómez-Goiri, A., Rodriguez-Gil, L., Diego, J., López-de-Ipiña D., & Garcia-Zubia, J. (2015). wCloud: Automatic generation of WebLab-Deusto deployments in the Cloud. In *Proceedings of 2015 12th International Conference on Remote Engineering and Virtual Instrumentation (REV)* (pp. 223–229).

Scapolla, A. M., Bagnasco, A., Ponta, D., & Parodi, G. (2005). A modular and extensible remote electronic laboratory. *International Journal of Online and Biomedical Engineering*, 1(1).

Part 2
Teaching with VISIR

Chapter 3

Experiments and Practices

3.1 Introduction

VISIR is a remote laboratory for electricity and electronics. Its main objective is to allow students to set up and measure different circuits as if they were in a hands-on laboratory. The characteristics of VISIR have already been described. In Part 2, we present different activities that can be performed with students in the classroom or at home.

This book differentiates between experiments and practices. In an experiment, the student tries to find a model (logical or mathematical) or to test it using an inquiry approach or not, respectively. For instance, with Ohm's law, students have two options: (1) students know Ohm's law and want to test it by setting up and measuring different circuits or (2) students set up and measure different circuits — offered by the teacher or not — to discover Ohm's law by themselves. For VISIR and for this book, it does not matter which strategy is selected by the teacher.

In a practice, students know the fundamentals of the circuits — Ohm's law, Kirchhoff's law, etc. — and what they are looking for is to use these fundamentals to set up and measure different concepts. In this situation, the main objective is simply to practice with circuits to learn and reinforce previously developed methods and abilities.

Part 2 mainly comprises experiments and practices. In general, in an experiment, we try to obtain or test a mathematical model, a graph (characteristic curve…), a data set, etc., but in a practice, we apply the knowledge obtained from the experiments in practical or specific situations. Each experiment or practice presents the following:

- an introductory section to explain the objective of the experiment/ practice;
- a model section to present the model or law that is discovered or tested in the experiment. If it is a practice, then maybe this section will be empty or it will include only the mathematical model to be used;
- an experimental/practical section with the circuits that will be set up and measured by the student;
- a conclusion section, if needed, to order and analyze the obtained results obtained in the previous section. In general, an experiment needs a conclusion section, while a practice does not need one.

The experiments and practices will be divided into five areas: DC circuits, AC circuits, and circuits with diodes, transistors, and operational amplifiers.

Before starting, readers should be aware of three ideas.

First, it is important to remark that the following set of experiments and practices are only a suggestion for readers and teachers; they are not the only way of using VISIR. Moreover, VISIR is an agnostic remote lab, that is, it never tells the user what should be done, or what comes next in an experiment, or what is wrong in a circuit; it simply offers the user devices and instruments.

Second, the practices and experiments that you find in the following sections need a particular configuration and installation of the VISIR. In this case, this is the configuration available at the University of Deusto. This configuration is not mandatory in general, of course, but it is needed to complete the following practices and experiments.

Finally, this book is not about electronics. The objective is not to teach electronics but to use VISIR with electronic circuits. The reader is assumed to be a teacher or someone with knowledge of circuits.

In Part 3, there is a discussion about the effect of remote experimentation in the learning process. This part is only a list of practices with VISIR.

3.1.1 *Configuration Files*

All the circuits explained in the following sections can be experimented and deployed in the VISIR remote laboratory, following the descriptions at the .max and components.list files.

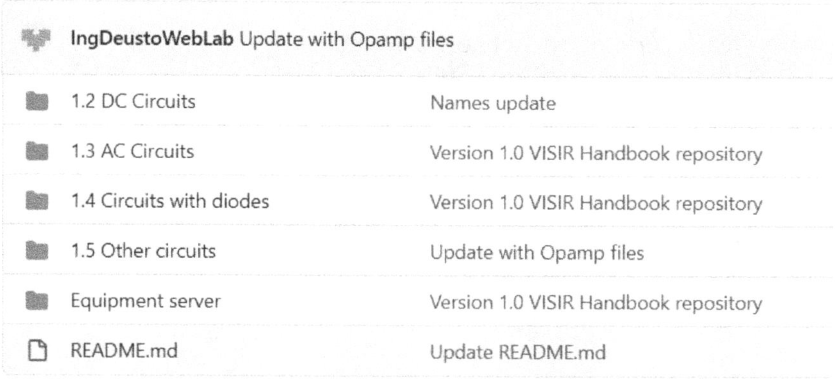

IngDeustoWebLab Update with Opamp files

1.2 DC Circuits	Names update
1.3 AC Circuits	Version 1.0 VISIR Handbook repository
1.4 Circuits with diodes	Version 1.0 VISIR Handbook repository
1.5 Other circuits	Update with Opamp files
Equipment server	Version 1.0 VISIR Handbook repository
README.md	Update README.md

Figure 3.1. GitHub structure.

Thus, all users of the remote laboratory, mainly lab technicians, who want to implement these experiments in their VISIR version, have all the necessary information to replicate them. Two configuration files are required to enable an experiment in VISIR:

- **.max files**: These files contain the description of the circuit that the user can implement on the VISIR client.
- **components.list file**: This file is unique for each VISIR version and describes the location of each component at the switching matrix.

For this purpose, a public and accessible repository has been created on GitHub: https://github.com/IngDeustoWebLab/VISIR_Handbook/. In this way, anyone can access it and download the files to execute the configuration defined.

The structure of this repository follows the outline of this chapter, as seen in Figure 3.1.

3.2 DC Circuits

The set of experiments and circuits is as follows:

- Experiment with resistors in series and parallel: Mathematical model
- DC Practice with resistors: Obtaining new resistances
- DC Practice with resistors: Can we measure the error?

- DC Experiment: Ohm's law
- DC Experiment: Kirchhoff's voltage law
- DC Experiment: Kirchhoff's current law
- DC Practice: DC power, voltages, and branches
- DC Practice: Voltage divider
- VISIR DC Power Source Experiment
- DC Experiment: Characteristic curve of a resistor
- DC Practice: Measuring DC circuits
- Thevenin and Norton Theorems
- Superposition Theorem.

.max files containing the descriptions of all the circuits included in the following subsections are included in the "1.2 DC Circuits" directory included at the GitHub repository. It is important to note that the Ohm_Kirchhoff_XX.max files are a set of files that together allow the user to build any circuit topology using up to 2 × 1 kΩ and 2 × 10 kΩ resistors simultaneously. In other words, they allow any combination/grouping of up to four resistors using at most 2 × 1 kΩ and 2 × 10 kΩ resistors simultaneously.

This combination of resistors has been chosen based on the didactic experience of the authors, and it offers a simple explanation of the implications of multiplying by 10 or dividing by 10 the value of a resistor in a branch of a circuit, with respect to the voltage or current flowing through it. Another pair of resistor values could be chosen by substituting in this set of .max files the values of 1 k and 10 k for the values selected by the teacher and updating the components.list accordingly.

3.2.1 *DC Experiment with Resistors in Series and Parallel: Mathematical Model*

1. *Introduction.* A resistor has a resistance measured in ohms (Ω) and is manufactured by companies to control the voltage and/or current in a circuit.

The resistors can be connected in series and parallel but ultimately, the whole connection can be considered one single resistor with a total resistance between A and B, where A and B used to be the two extremes — left and right — of the connection.

2. *Mathematical model to be discovered or tested.* If two or more resistors are connected in series, the total resistance is

$$R_{\text{tot}} = R_1 + R_1 + R_3 + \cdots + R_n$$

If two or more resistors are connected in parallel, the total resistance is

$$\frac{1}{R_{\text{tot}}} = \frac{1}{R_1} + \frac{1}{R_2} + \frac{1}{R_3} + \cdots + \frac{1}{R_n} \quad \text{for only } R_1 \text{ and } R_2, R_{\text{tot}} = \frac{R_1 \cdot R_2}{R_1 + R_2}$$

3. *Experiment.* In VISIR, there are two 1 k resistors and two 10 k resistors. They can be connected in any way. Table 3.1 shows three connections: series, parallel, and series–parallel.

Table 3.1. Resistors, connections, and measurements.

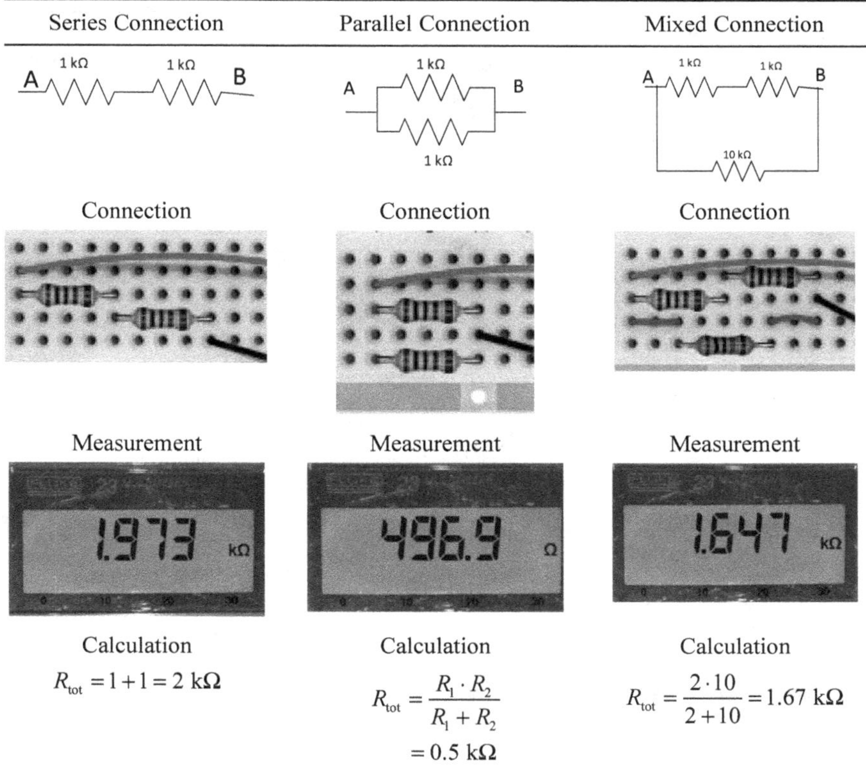

Series Connection	Parallel Connection	Mixed Connection
A 1 kΩ 1 kΩ B	A 1 kΩ 1 kΩ B	A 1 kΩ 1 kΩ 10 kΩ B
Connection	Connection	Connection
Measurement	Measurement	Measurement
1.973 kΩ	496.9 Ω	1.647 kΩ
Calculation	Calculation	Calculation
$R_{\text{tot}} = 1 + 1 = 2\ \text{k}\Omega$	$R_{\text{tot}} = \frac{R_1 \cdot R_2}{R_1 + R_2} = 0.5\ \text{k}\Omega$	$R_{\text{tot}} = \frac{2 \cdot 10}{2 + 10} = 1.67\ \text{k}\Omega$

4. *Analysis and conclusions.* After setting up and measuring different connections of resistors, the mathematical model can be said to be correct: All the measurements obtained are similar to the calculated values. They are not equal because when using real devices and instruments, there are intrinsic errors.

5. *Special connections.* Two special connections are interesting: what happens if a short circuit is connected in series or in parallel? What happens if the two connections are with an open circuit?

A short circuit is a wire with 0-Ω resistance and an open circuit is a broken circuit or an absence of wire with ∞-Ω resistance.

Table 3.2 shows and measures the four connections.

Table 3.2. Special connections with resistors.

Circuit	Measure	Conclusion
Short circuit in series 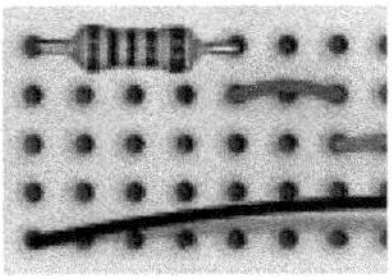	 992.5 Ω	It does not matter, you can do it.
Open circuit in series 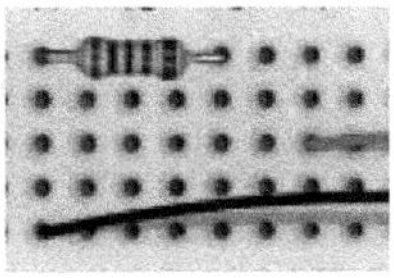	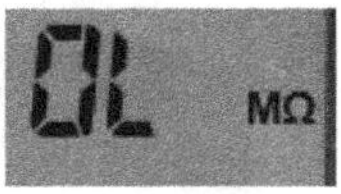 OL MΩ	It matters, do not do it. Over limit in MΩ.
Open circuit in parallel 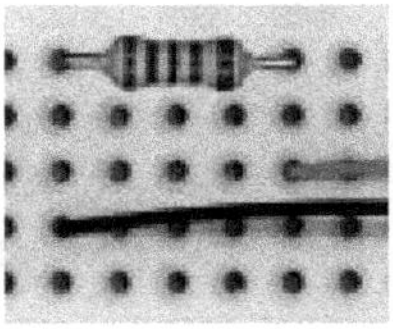	 990.7 Ω	It does not matter, you can do it.
Short circuit in parallel 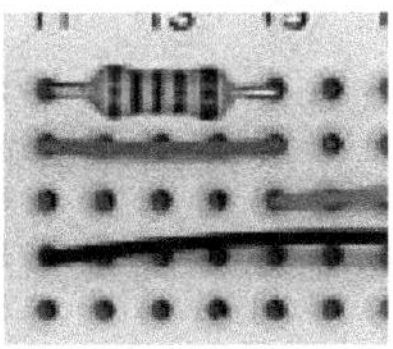	 170.2 mΩ	It matters, do not do it. 170 mΩ is close to 0 Ω.

The use of the mathematical model reinforces the previous experiment.

3.2.2 *DC Practice with Resistors: Obtaining New Resistances*

1. *Introduction.* After the previous experiments, we know how to connect resistors and measure resistance. Now we can use this prior knowledge in a new scenario.

In VISIR, we can create any series–parallel connection with four resistors: 2 × 1 kΩ and 2 × 10 kΩ, but if we need another resistance, can we obtain it?

2. *Mathematical model to be used.* If two or more resistors are connected in series, the total resistance is

$$R_{\text{tot}} = R_1 + R_1 + R_3 + \cdots + R_n$$

If two or more resistors are connected in parallel, the total resistance is

$$\frac{1}{R_{\text{tot}}} = \frac{1}{R_1} + \frac{1}{R_2} + \frac{1}{R_3} + \cdots + \frac{1}{R_n} \quad \text{for only } R_1 \text{ and } R_2, R_{\text{tot}} = \frac{R_1 \cdot R_2}{R_1 + R_2}$$

3. *Practice.* The question is, for instance, whether it is possible to obtain a total resistance of approximately 1.8 kΩ with the VISIR set of resistors. The answer is yes.

Look at Figure 3.2 and note that a series connection of two parallel connections of 1 kΩ and 10 kΩ results in this value.

In Figure 3.3, we can see how to obtain a value of 6 kΩ of resistance.

Can we obtain a resistance value between 0.4 kΩ and 0.5 kΩ? And around 0.86 kΩ?

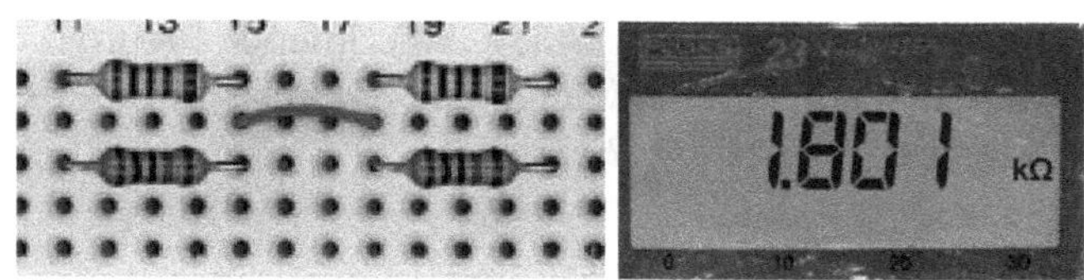

Figure 3.2. Resistors connection I.

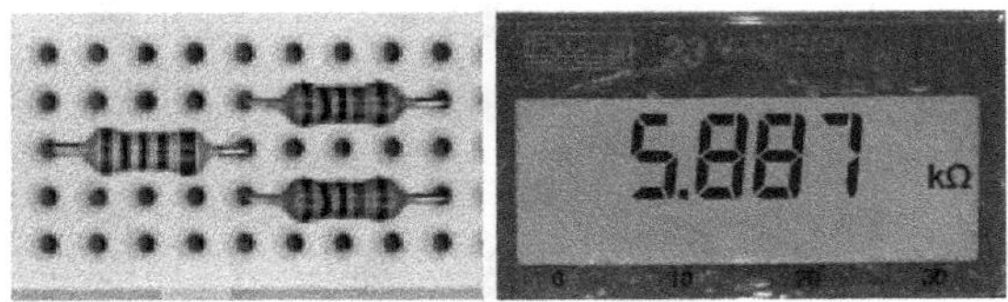

Figure 3.3. Resistors connection II.

4. *Conclusion.* Connecting resistors in series and parallel, we can obtain new resistance values.

3.2.3 *DC Practice with Resistors: Can We Measure the Error?*

1. *Introduction.* As we have seen in the previous practice, by connecting 1-kΩ and 10-kΩ resistors, we obtain new resistance values. There are two new values: the one expected from the mathematical model and the real value measured in the VISIR. The most important one is the measured value because it is real, but why is the measured value different from the calculated value? The answer is because there are errors and nature is not familiar with mathematical models. In the last example, the resistance expected value was 6 kΩ, but the real measured value was 5.887 kΩ. What happens?

For instance, when we buy a 1-kΩ resistor, its value is more or less 1 kΩ; it is not exact, as it is when we buy one liter of milk. This is one source of error. Also, when we include a wire, its ideal resistance is 0 Ω, but this is not real; it is a small value, but it is not 0. And of course, the multimeter tries to do its best, but it is not ideal, it is real. To measure is to make errors, unfortunately.

2. *Practice*. Thus, when we measure the resistor or the connection of resistors, we are measuring different things. The deviation, the difference between the expected value and the real value, is not only produced by the multimeter, but also in other parts of the experiments. Figure 3.4 shows a connection of two 10k in parallel with another 1k resistor in series (in total 6 kΩ); the measure value is 5.932 Ω.

This deviation is associated with the VISIR itself (relays, contacts, boards…), with the tolerance of the resistor, with room temperature, which affects resistance, etc. The resistors included in VISIR at Deusto are "golden resistors", i.e., their tolerance is 5%, so the result should be between 5700 Ω and 6300 Ω, so there is not a big error in this measurement.

If we measure a short circuit with the ohmmeter, we can see the value given in Figure 3.5.

3. *Conclusion.* Mathematical models and real circuits are not exactly the same. It is important to understand that real circuits are not ideal, but they are close to it. The use of mathematical models helps us to find the best solution for an electrical problem. However, after the calculation, we need to set up and measure the real circuit.

And remember — to measure is to make mistakes.

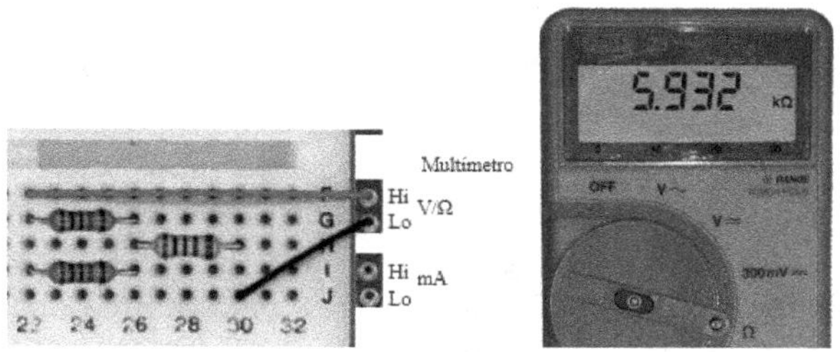

Figure 3.4. Resistors connection III.

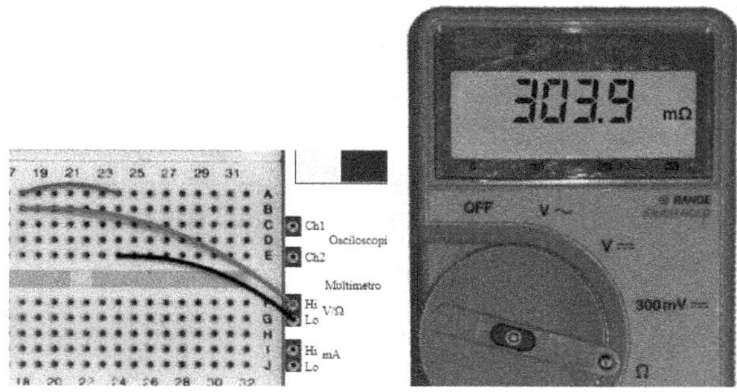

Figure 3.5. Resistors connection IV.

3.2.4 *DC Experiment: Ohm's Law*

1. *Introduction.* An electric circuit can be described with three different dimensions or variables: resistance, voltage, and current. Between two points (A and B) of an electric circuit, there is a resistance, a voltage drop, and an electric current. Between these two points, there is normally at least one resistor (this could be another device, not only a resistor). If the reference is only one point in the circuit, this means that all the measurements are between this point and ground, except the current that flows in the wire.

In the 19th century, George Ohm (1789–1854, Germany) declared that these three signals are interrelated.

2. *Mathematical model.* Ohm's law says the following:

$$V_{\mathrm{AB}} = R_{\mathrm{AB}} \cdot I_{\mathrm{AB}}, \quad V = I \cdot R, \; I = \frac{V}{R}, \quad R = \frac{V}{I}$$

This means that there is a lineal relation between these three signals, and if two are known, the third can be calculated using the previous equation.

3. *Experiment.* The basic idea is to set up different circuits with resistors powered by a known DC voltage, measure the current between them, and then test whether or not Ohm's law and its equation are correct.

Figure 3.6 shows how a circuit is powered with 6 VDC in VISIR.

To measure the resistance, we have to remove the power, otherwise the measurement will not be correct (Figure 3.7).

To measure the voltage, we do not need to open the circuit, as the multimeter is in parallel. The voltage is measured between the + terminal and the – terminal of the multimeter. If wires are exchanged, the measurement will be negative (Figure 3.8). The voltage is close to 6 V.

To measure the current, we need to open the circuit and place the multimeter in series with the circuit, not in parallel as it was for the voltage. It is also necessary to use the current terminals instead of the voltage/resistance terminals (Figure 3.9).

Thus, the measurements are V = 5.998 V, R = 0.995 kΩ, and I = 6.077 mA.

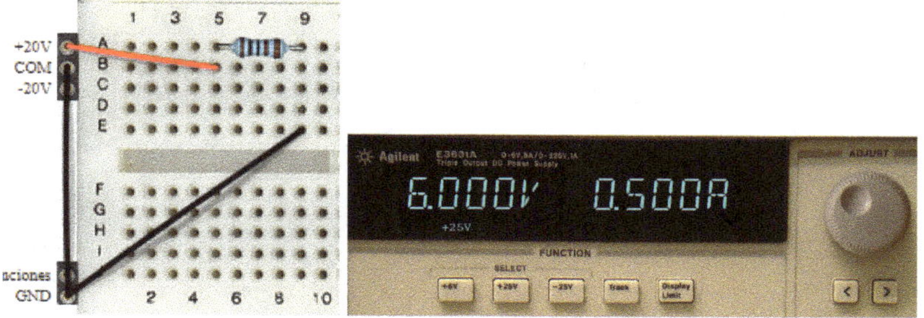

Figure 3.6. DC Circuit: Ohm's Law.

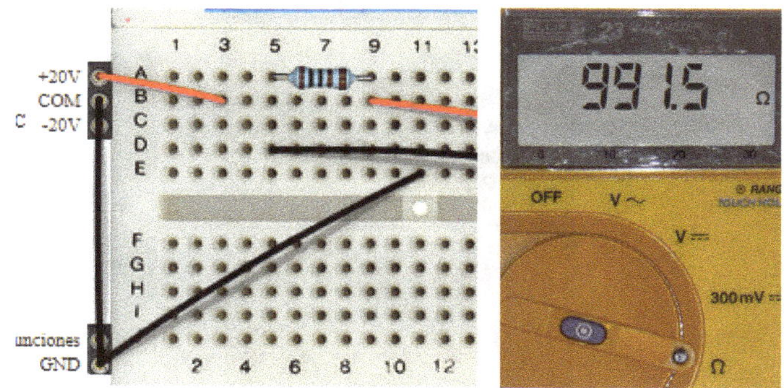

Figure 3.7. DC Circuit: Resistance measurement.

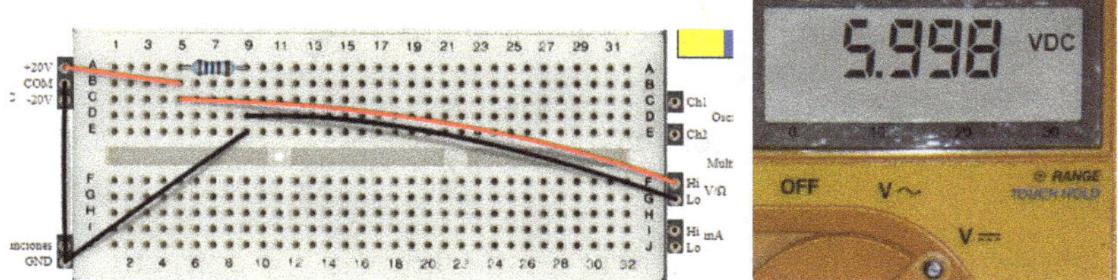

Figure 3.8. DC Circuit: Voltage measurement.

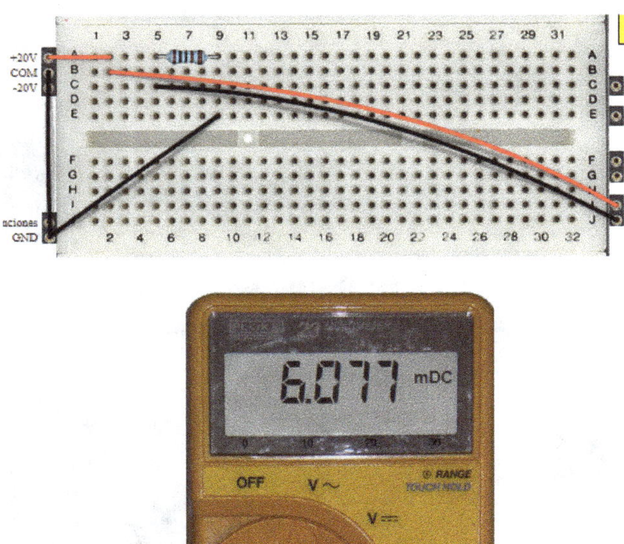

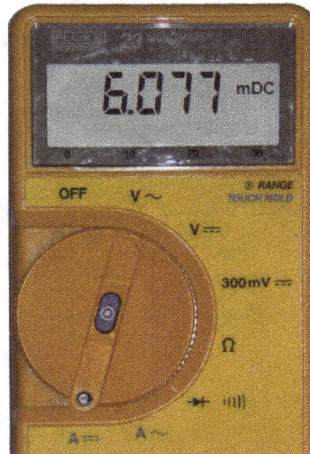

Figure 3.9. DC Circuit: Current measurement.

The expected value for the current is 6 mA,

$$I = \frac{V}{R} = \frac{6}{1} = 6\ \text{mA} \quad \text{or maybe} \quad I = \frac{V}{R} = \frac{5.998}{0.995} = 6.028\ \text{mA}$$

Therefore, in this experiment at least, Ohm's law is correct.

Now we should repeat the experiment with different voltages and resistors. Table 3.3 shows three additional experiments.

All the examples of Experiment 3.2.1 can be powered and can be measured to test Ohm's law.

Ohm's law is not only true for the total current. It is also correct for any point of the circuit. If we measure the current in the 10-kΩ branch in the third circuit of Table 3.3, we obtain Figure 3.10.

Table 3.3. DC examples.

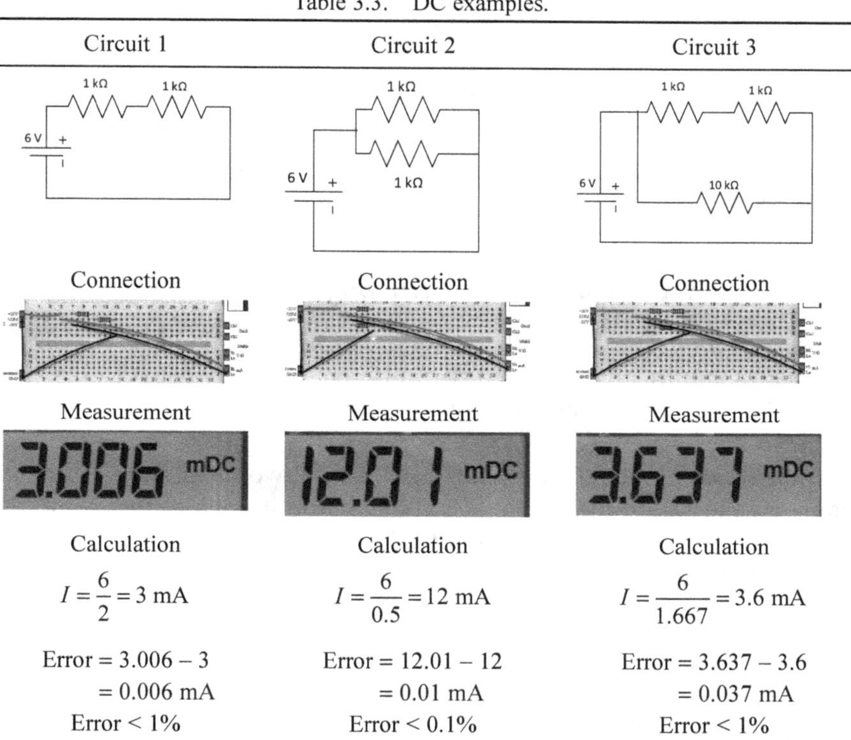

Circuit 1	Circuit 2	Circuit 3
1 kΩ, 1 kΩ, 6 V	1 kΩ, 1 kΩ, 6 V	1 kΩ, 1 kΩ, 10 kΩ, 6 V
Connection	Connection	Connection
Measurement	Measurement	Measurement
3.006 mDC	12.01 mDC	3.637 mDC
Calculation	Calculation	Calculation
$I = \frac{6}{2} = 3$ mA	$I = \frac{6}{0.5} = 12$ mA	$I = \frac{6}{1.667} = 3.6$ mA
Error = 3.006 − 3 = 0.006 mA	Error = 12.01 − 12 = 0.01 mA	Error = 3.637 − 3.6 = 0.037 mA
Error < 1%	Error < 0.1%	Error < 1%

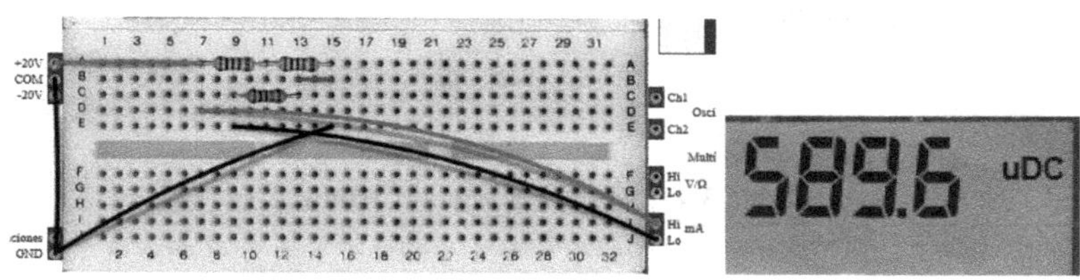

Figure 3.10. DC: Current measurement in the 10-kΩ branch.

In this case, Ohm's law produces a value of 0.6 mA, i.e., 600 μA, and the measurement is 589.6 μA. The error is around 10 μA, less than 2%.

$$I = \frac{V}{R} = \frac{6}{10} = 0.6\,\text{mA}$$

4. *Analysis and conclusion.* There is a linear relation between *V*, *I*, and *R*. If we know or measure two of these values, we can obtain the other using Ohm's law.

3.2.5 *DC Experiment: Voltage Kirchhoff's Law*

1. *Introduction.* In the 19th century, Gustav Kirchhoff (1824–1887, Germany) empirically observed two different relations between resistors and voltage/current. There are two of Kirchhoff's laws.

2. *Mathematical model.* Kirchhoff says that if two or more resistors are connected in series, then there is a voltage drop in each of them, and the sum of all of them must be equal to the DC power. This is Kirchhoff's second law

$$V_{\text{DC}} = V_{R1} + V_{R2} + V_{R3} + \cdots + V_{Rn}$$

3. *The experiment.* Three circuits, with two, three, and four resistors in series, are set up, and the voltage drop in each resistor is measured (Table 3.4). The DC power is 6 V.

Table 3.4. Voltage Kirchhoff's Law.

Circuit 1	Circuit 2	Circuit 3

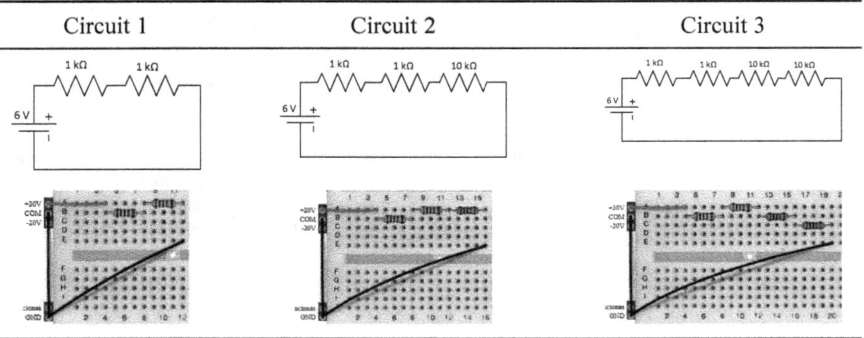

Table 3.4. (*Continued*)

Circuit 1	Circuit 2	Circuit 3
3.023 VDC	502.5 mVDC	275.2 mVDC
2.981 VDC	502.7 mVDC	271.9 mVDC
$V_{tot} = 6.004$ V	4.996 VDC	2.747 VDC
	$V_{tot} = 6.001$ V	2.709 VDC
		$V_{tot} = 6.003$ V

4. *Analysis and conclusion.* Kirchhoff's second law is correct. That is, all the DC power in a circuit with resistors in series is divided among all the resistors. The voltage drop in each resistor is related to the resistance, as Ohm's law establishes.

3.2.6 *DC Experiment: Current Kirchhoff's Law*

1. *Introduction.* In the 19th century, Gustav Kirchhoff (1824–1887, Germany) empirically observed two different relations between resistors and voltage/current. There are two Kirchhoff's laws.

2. *Mathematical model.* Kirchhoff says that if two or more resistors are connected in parallel, then the total current is divided among all the branches. The sum of all of the currents must be equal to the total current. This is Kirchhoff's first law:

$$I_{\text{tot}} = I_{R1} + I_{R2} + I_{R3} \cdots + I_{Rn}$$

3. *The experiment.* Three circuits, with two, three, and four resistors in parallel, are set up and the current in each resistor is measured. The DC power is 6 V.

At this point, it is important to explain how we can measure the current in a branch. There are many different ways of doing this, but there is one that is simple to apply: Move the resistor to a new place, connect the

the wire from the HI channel of the multimeter at the point of the first terminal of the resistor, connect the wire from LO to the first terminal of the resistor (wire 3 in the new position), and finally add a wire from the second terminal of the resistor to the original position of the second terminal of the resistor (step 4 in Figure 3.11).

Table 3.5 sums up the measurements of three different circuits with resistors in parallel.

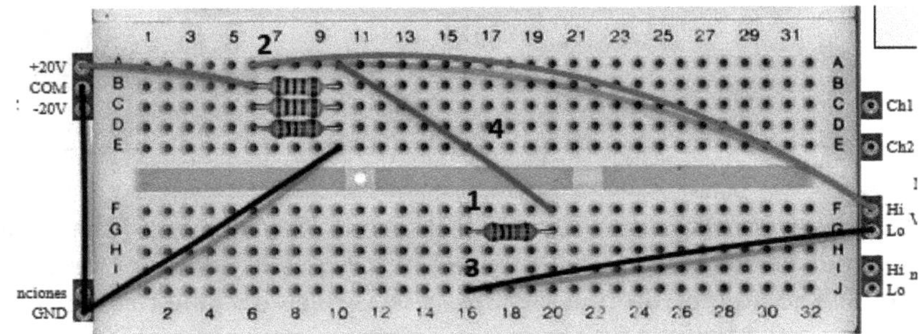

Figure 3.11. Current measurement method.

Table 3.5. Current measurements in DC circuits.

Circuit 1	Circuit 2	Circuit 3
6 V, 1 kΩ, 1 kΩ	6 V, 1 kΩ, 1 kΩ, 10 kΩ	6 V, 1 kΩ, 1 kΩ, 10 kΩ

Table 3.5. (*Continued*)

Circuit 1	Circuit 2	Circuit 3
12.03 mDC	12.72 mDC	13.20 mDC
6.040 mDC	6.042 mDC	6.047 mDC
6.033 mDC	6.040 mDC	6.055 mDC
I_{tot} = 12.073 mA	595.5 uDC	595.7 uDC
	I_{tot} = 12.678 mA	595.9 uDC
		I_{tot} = 13.294 mA

Looking at the results, we again see Ohm's law: The higher the resistance, the lower the current. For instance, if there is a current of 6 mA in a 1-kΩ resistor, the current in a 10-kΩ resistor will be 0.6 mA.

It is important to note that in the previous examples, at any time that we modify the circuit (adding new resistors in parallel), then the total resistance is modified, and because of that, the total current is different. The first value in each column of Table 3.5 is the total current available in that circuit.

4. *Analysis and conclusion.* Kirchhoff's first (current) law is correct. That is, all the DC (total current) in a circuit with resistors in parallel is split among all the branches. The voltage drop in each resistor is related to the resistance, as Ohm's law establishes.

3.2.7 *DC Practice: DC Power, Voltages, and Branches*

1. *Introduction.* If a circuit is powered with a DC power supply, then the voltage is expected to be the same in any circuit. That is, if we have a

circuit powered with 6 V and we add new resistors, the total voltage available will be 6 V. This was observed in Experiment 1.2.5.

However, what happens with the current?

2. *Logical and mathematical model.* Logically, if we have a circuit with 1 kΩ and powered with 6 V, then the current is 6 mA. If we add a new 1-kΩ resistor, we might reason as follows: "OK, now there will be 3 mA for the first resistor, and the same for the new one". That is, the total current is split into two identical branches. Is this correct?

At the same time, however, Ohm's law can be stated as follows:

$$I = \frac{V}{R}$$

Thus, adding a new resistor in parallel modifies the circuit and its total resistance, so this modifies the total current. What happens?

3. *Experiment.* In this case, we do not need to perform a new experiment; all we need is to consult Table 3.5.

In each column of Table 3.5, there is a different circuit: The first had two resistors, the second three resistors, and the third our resistors. What was the behavior of the current? At any time, the current in a resistor was the same, even in different circuits.

In the first circuit, each 1-kΩ resistor had 6 mA, then we added a 10-kΩ resistor, and the current was the same, 6 mA, while the new resistor had 0.6 mA. Finally, we added another 10-kΩ resistor, and all the previous measurements remained the same.

It may sound a little magical but looking at Table 3.5, every time we added a new resistor, the total current grew to attend the current demand of the new branch.

In this case, the power of the DC source increases, i.e., we have to pay more, or if we use batteries, they will discharge quickly. Thus, in a DC voltage source, the voltage is fixed and the current increases or decreases depending on the connected resistors.

What happens in the reverse situation? If we add resistors in series, the current decreases for all the resistors. This is because we use a DC voltage source, but we can use a DC source. With this DC source, the current is always the same, and we modify the circuit; the voltage will then be automatically modified by the DC source.

4. *Analysis and conclusion.* The conclusion is very simple and useful: We can add resistors (or other devices) in parallel and the current in the original resistors will not be affected. If we think of our homes, we can see that all the devices are connected in parallel.

3.2.8 *DC Practice: Voltage Divider*

1. *Introduction.* Imagine that we have a DC power source with a fixed value, 6 V for instance, but need to have 3 V at the circuit output. Can this value be obtained?

2. *Mathematical model.* In Experiment 1.2.5, we saw that if we add a resistor in series to a previous circuit, then the voltage and the current are affected.

3. *Practice.* If a circuit is powered with 6 V, can we obtain 3 V at the output? Figure 3.12 shows the solution to this problem.

Thus, regarding the previous circuit, there are infinite possible solutions to the problem; R_o must simply be equal to R. For instance, $R = R_o = 1\ \text{k}\Omega$.

Obviously, however, $R = R_o = 10\ \text{k}\Omega$ is also a solution (Figure 3.13).

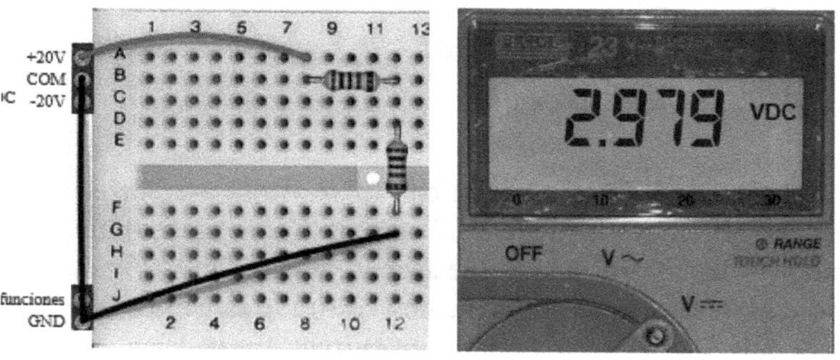

Figure 3.12. Voltage divider I.

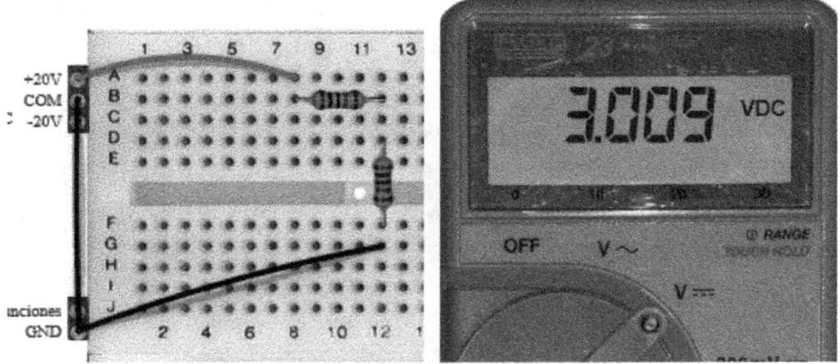

Figure 3.13. Voltage divider II.

Figure 3.14. Voltage dividers' current measurements.

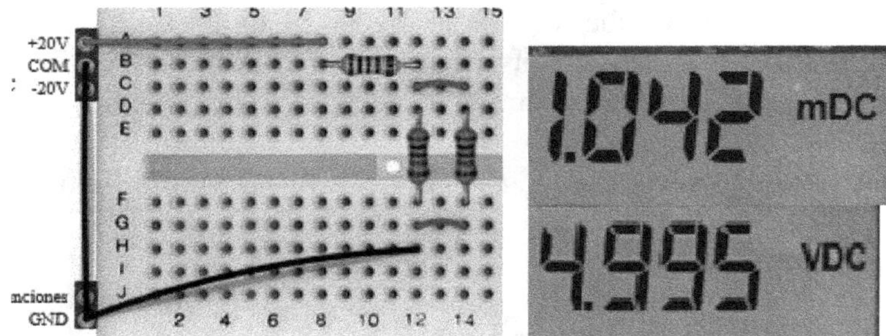

Figure 3.15. Voltage divider III.

If the two previous combinations of resistors are correct, which is better? The difference is in the total current. In the first solution, the total current is around 3 mA, and in the second, it is 0.3 mA (Figure 3.14).

If the current is lower, so is the power. At the same time, however, we want not only 3 V but also a minimum current, for instance, 1 mA, so the $R = R_0 = 1\ \mathrm{k}\Omega$ is better. It depends on the design requirements.

If the requirement is to obtain 5 V at the output with a DC power source of 6 V, the solution is Figure 3.15. The output is 5 V and the current is 1 mA.

In this design, we combine resistors to obtain a 5-kΩ resistor, as we learned in a previous practice.

4. *Analysis and conclusion.* We have seen that we can modify a circuit to obtain a determined requirement. That is, we can design a solution and test it in the VISIR.

The voltage divider simply offers the designer a means of obtaining a specific voltage at the output.

3.2.9 *VISIR DC Power Source Experiment*

1. *Introduction.* VISIR limits the current to avoid problems. If the current increases, the equipment or the circuits under experimentation (CUT) could be destroyed and there is even a risk of fire.

What is the maximum current allowed by VISIR? What happens if we try to exceed this maximum current?

2. *Mathematical model.* We need to remember Ohm's law:

$$I = \frac{V}{R}$$

3. *Experiment.* According to Ohm's law, we should create a basic circuit and then increase the voltage to see whether or not the current is limited. Before constructing the circuit, we need to remember that to reach a "high" current, we need to power the circuit with a "high" voltage or use a resistor with a low resistance. Figure 3.16 shows a circuit with a 6-Ω resistor. This resistor supports 25 W.

Table 3.6 shows the behavior of the circuit.

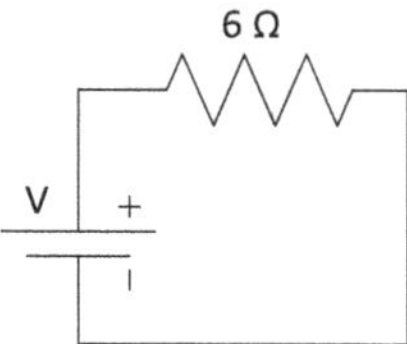

Figure 3.16. Power source experiment with a simple DC circuit.

Table 3.6. Measurements of a DC circuit.

$V = 0.2$ V	$V = 0.4$ V	$V = 0.6$ V	$V = 0.8$ V	$V = 1$ V	$V = 1.2$ V
24.3 mA	48.7 mA	73.0 mA	97.2 mA	99.6 mA	99.6 mA
0.2 V	0.4 V	0.6 V	0.8 V	0.82 V	0.82 V

Figure 3.17. Current limitation in the VISIR DC power source.

4. *Analysis and conclusion.* In the first row of Table 3.6, we see the input voltage, and in the second row, we see the current. We can see that if we increase the voltage, the current does not increase; it is limited to around 100 mA. This is the current limit in VISIR at Deusto.

In the third row, we see the voltage of DC power source after "perform experiment". That is, VISIR powers the circuit from 0 V to 1 V, but when the current exceeds the limit, then the voltage is cut to guarantee this current. Figure 3.17 shows that the voltage is cut when the current limit is exceeded. One can also see on the interface that the current limit is 0.5 A, but this is incorrect.

Looking again at Table 3.6, we see that with 0.6 V the current is not 100 mA, as was expected using Ohm's law. At the same time, however, and using Ohm's law with the data obtained from 0 V to 0.8 V, the resistance obtained is 8.2 Ω, instead of the expected 6 Ω. Why?

Why are there 2.2 additional ohms? There could be different reasons, but we must remember that the circuit is set up in VISIR using relays, and each relay implies a low-value new resistor. In this case, it is 2.2 Ω.

3.2.10 *DC Experiment: Characteristic Curve of a Resistor*

1. *Introduction.* Every device has a characteristic curve. This curve relates the voltage drop and the current at the selected device. In our case, the device is a resistor.

Companies provide clients with the voltage–current characteristic to explain to them what are they buying.

This experiment could be performed just after Experiment 1.2.4 (Ohm's law) to explain the relation between voltage, current, and resistance.

2. *Mathematical.* Ohm's law specifically describes the relation between V, I, and R:

$$V = I \cdot R, \quad R = \frac{V}{I}$$

Graphically, this means that if we draw V in the X-axis and I in the Y-axis, then the slope of this curve is R, the resistance.

3. *Experiment.* First, we select a resistor, 1 kΩ, for instance. The experiment is very simple: From –2 V to +2 V with a step of 0.5 V measuring the voltage and the current (Table 3.7).

To obtain a negative voltage, we need to use the –20-V DC power source. Remember to select this value in the DC power interface (Figure 3.18).

The slope in Figure 3.19 is 1 (more or less). That is, the resistance is 1 kΩ, because the voltage is in V, and the current is in mA.

This experiment can be repeated with 10 kΩ, 100 Ω, etc.

Table 3.7. Characteristic curve data.

	–2 V	–1.5 V	–1 V	–0.5 V	0 V	0.5 V	1 V	1.5 V	2 V
V	–1.996	–1.497	–0.999	–0.499	0.001	0.499	0.999	1.501	2.001
I	–2.035	–1.531	–1.024	–0.518	0.0001	0.528	1.034	1.536	2.043

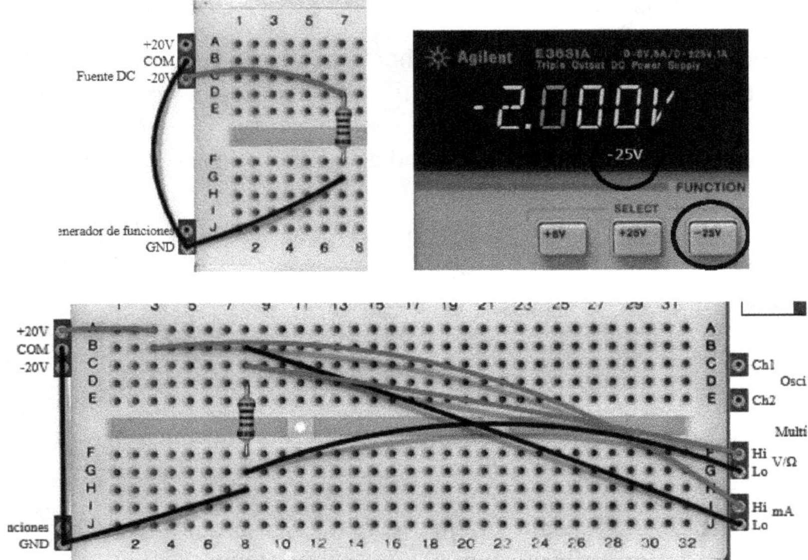

Figure 3.18. Resistor characteristic curve obtention.

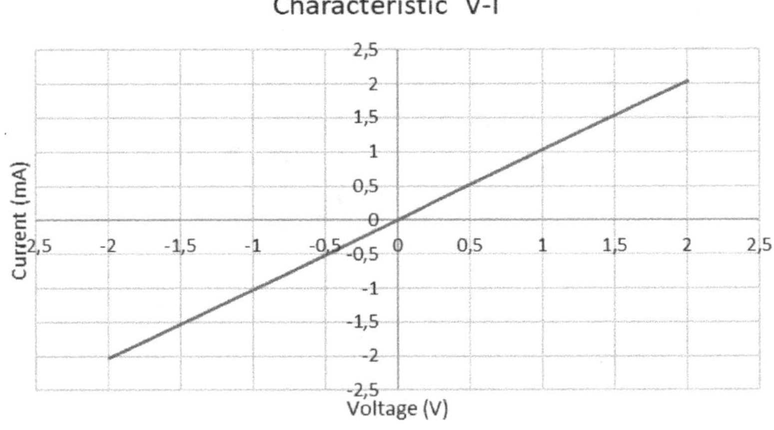

Figure 3.19. Resistor characteristic curve with data from Table 3.7.

4. *Analysis and conclusion.* After the experiment, resistance can be said to be constant for every voltage and current, and its value is the slope of the *V–I* characteristic or *V*/*I*. In reality, one can say that this resistor has a resistance of at least 1 kΩ, at least from –2 V to + 2 V.

We should repeat this for every resistor and for a wider voltage range, or we can trust the resistor manufacturers.

3.2.11 *DC Practice: Measuring DC Circuits*

1. *Introduction.* Until now, we have used VISIR for testing some laws and theorems and for analyzing some practical applications of the latter. However, VISIR can be used to set up and measure any DC circuit. That is, we solve a DC circuit on paper/blackboard with pencil/chalk, and then we check whether or not the obtained results are correct.

The meaning of the sentence "any DC circuit" should be any DC circuit that can be constructed with our VISIR configuration.

2. *Mathematical model.* To solve a DC circuit, we need only Ohm's law and Kirchhoff's laws:

$$V = I \cdot R, \quad V_{\text{tot}} = V_1 + V_2 + \cdots + V_n, \quad \text{and} \quad I_{\text{tot}} = I_1 + I_2 + \cdots + I_n$$

Applying these equations, the values of all the signals of a DC circuit (Figure 3.20) can be obtained (Table 3.8). Each resistor has a voltage drop and an intensity current flowing through it.

3. *Experiment.* The previous circuit can be set up and measured in VISIR, as shown in Figure 3.21.

Now we must measure all the signals/variables using the knowledge acquired in the previous experiments.

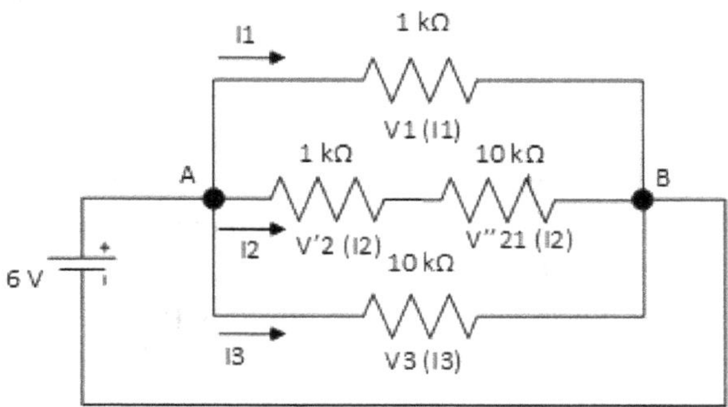

Figure 3.20. DC circuit under experimentation.

Table 3.8. Mathematical results of the DC circuit under experimentation.

R_{tot} (kΩ)	I_{tot} (mA)	I_1 (mA)	I_2 (mA)	I_3 (mA)	V_1 (V)	V_2' (V)	V_2'' (V)	V_3 (V)
0.84	7.14	6	0.55	0.6	6	0.545	5.45	6

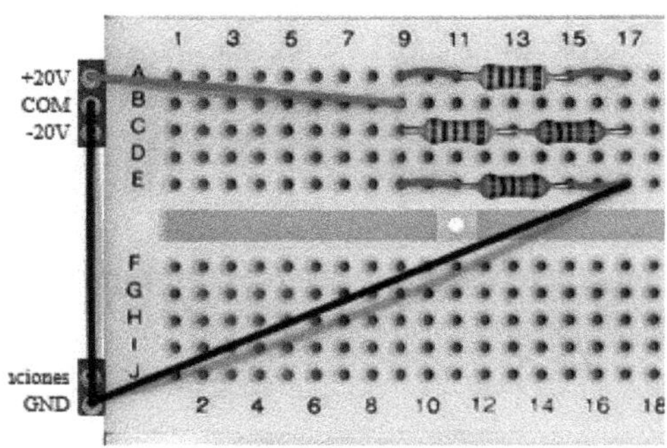

Figure 3.21. DC circuit under experimentation in VISIR.

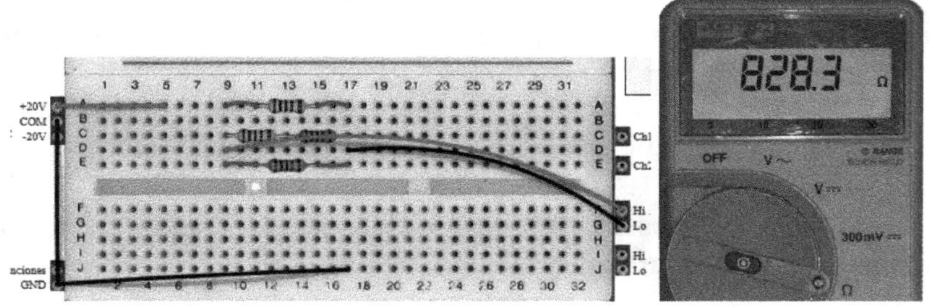

Figure 3.22. Resistance measurement in the DC circuit under experimentation in VISIR.

To measure the total resistance, the circuit must not be powered (Figure 3.22).

To measure the voltage drop, the circuit must be powered and the circuit must not be modified. The voltage drop in the second 1-kΩ resistor (subject to I_2) is shown in Figure 3.23.

To measure the current, the circuit must be powered and the circuit must be open. The total current of the circuit is shown in Figure 3.24.

Instead of measuring the total current of the circuit where it begins, we can measure the current at the end (Figure 3.25).

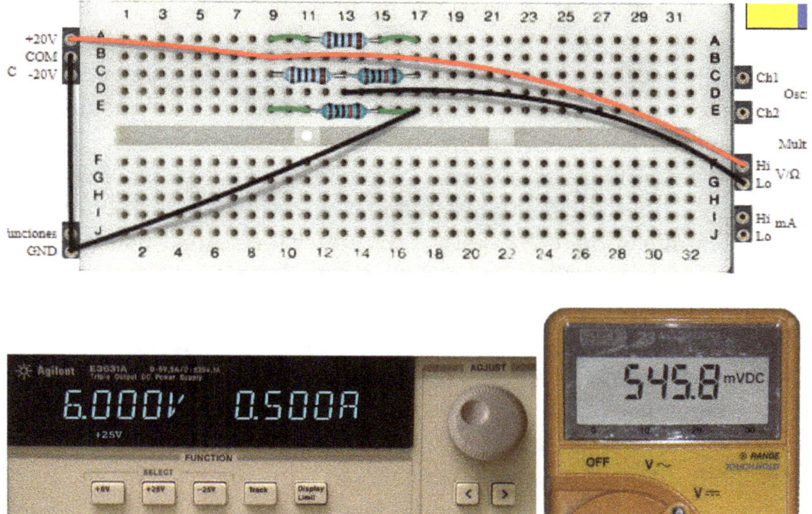

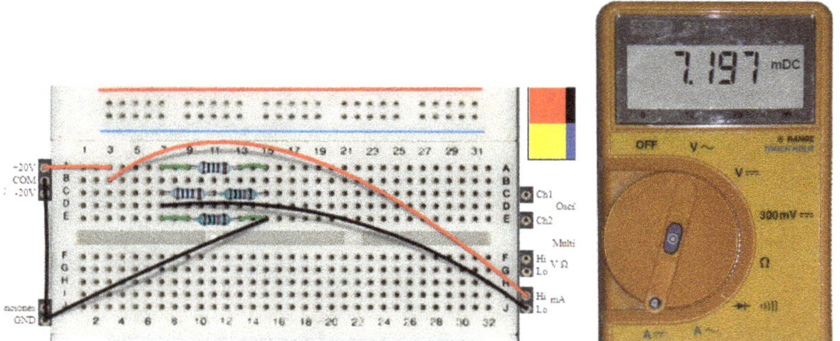

Figure 3.23. Voltage measurement in the DC circuit under experimentation in VISIR.

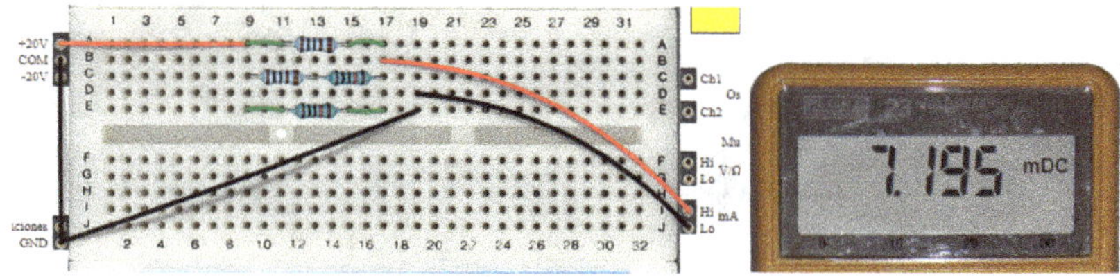

Figure 3.24. Current measurement in the DC circuit under experimentation in VISIR.

Figure 3.25. Different current measurement in the DC circuit under experimentation in VISIR.

One thing, however, should not be forgotten: measuring the circuit current in VISIR requires resources. That is, to measure the current at the beginning and at the end of a circuit, double resources are needed, compared to measuring in only one position. And VISIR lab resources are limited. That is, when configuring the VISIR remote lab, a decision must be taken: measure the current in all possible paths but reduce the number of circuits, or allow the maximum number of circuits but reduce the ways of measuring the current. At Deusto, we prefer the second option.

To measure the current in any branch, the VISIR user needs to make an extra effort. One strategy is to remove the affected resistor from the circuit (Figure 3.26). The current in the first branch is measured.

However, we can also measure the current using another simple strategy: add wires before all the branches. In this circuit, we have wires in two of the branches (first and third in Figure 3.27). Before measuring, look at the third wire because it needs to be moved.

If we want to follow this second strategy, the original circuit should be as follows. In this case, to measure the current we remove the wire and connect the multimeter between columns 9 and 10 (Figure 3.28).

Table 3.9 shows all the circuit measurements.

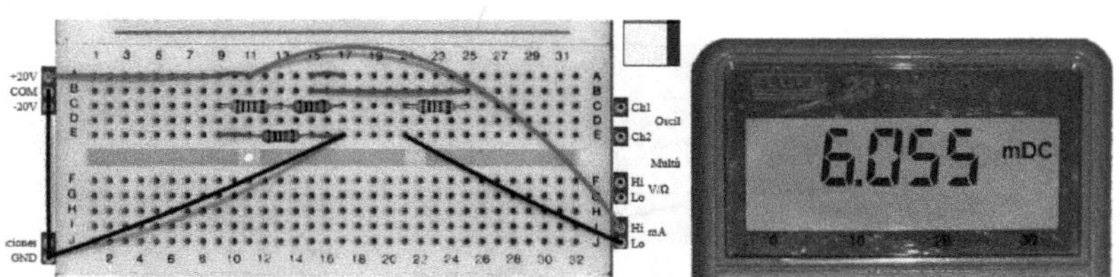

Figure 3.26. Current measurement in branch 1 of the DC circuit under experimentation in VISIR.

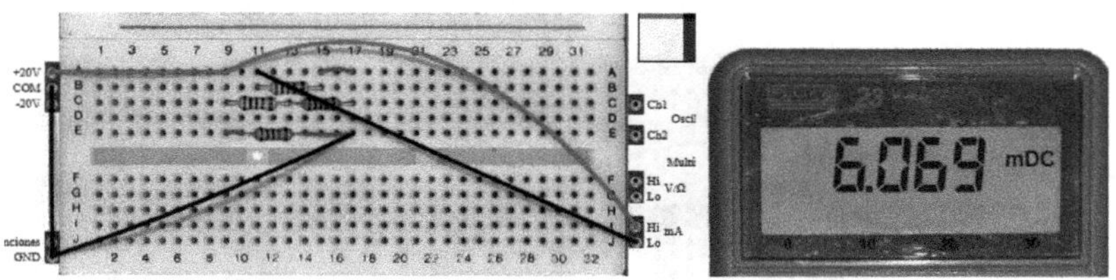

Figure 3.27. Current measurement in branch 2 of the DC circuit under experimentation in VISIR.

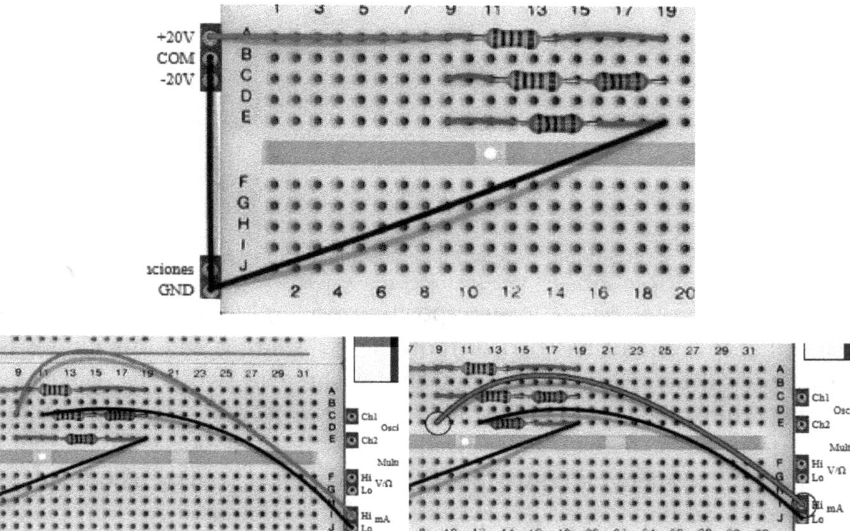

Figure 3.28. Different strategy for current measurement in the DC circuit under experimentation in VISIR.

Table 3.9. Summary of the DC circuit under experimentation.

R_{tot} (kΩ)	I_{tot} (mA)	I_1 (mA)	I_2 (mA)	I_3 (mA)	V_1 (V)	V_2' (V)	V_2'' (V)	V_3 (V)
0.828	7.197	6.055	0.5327	0.5963	5.996	0.5458	5.450	5.996

4. *Analysis and conclusion.* The analytical results and the empirical results must be equal or at least similar.

3.2.12 *DC Experiment: Superposition Principle*

1. *Introduction.* If a circuit is powered by two (or more) power sources, what can be done in VISIR?

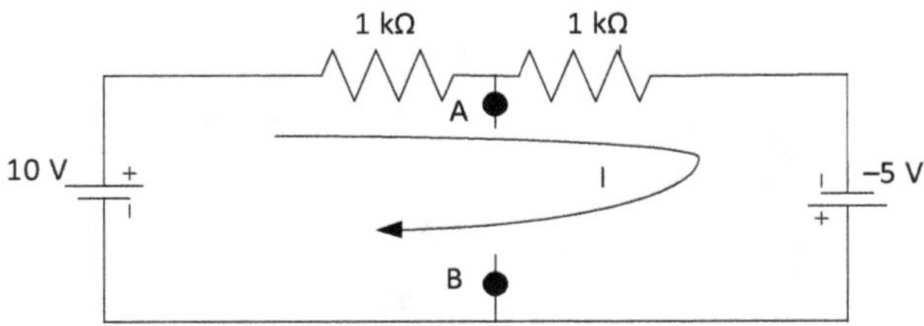

Figure 3.29. Circuit under experimentation for the superposition principle.

2. *Mathematical model.* The superposition principle (or theorem) establishes that if a circuit has two or more power sources and this circuit is linear, then the voltage/current can be measured in two steps. First, we calculate the voltage/current for one of the power sources, while the other is short-circuited, and then we short-circuit the first power source and calculate the voltage/current with the second power source. The result is obtained by adding the results obtained in each calculus.

Figure 3.29 is a circuit with two power sources: +10 V and –5 V. The calculi are as follows:

$$\text{For } +10\text{V}, \quad R_{\text{tot}} = 1 + 0.91 = 1.91 \quad \text{and} \quad I_{\text{tot}} = \frac{10}{1.91} = 5.24\,\text{mA},$$

$$\text{so } I_{10} = 5.24\frac{1}{11} = 0.48\,\text{mA} \quad \text{and} \quad V_{10} = 4.8\,\text{V}$$

$$\text{For } -5\text{V}, \quad R_{\text{tot}} = 1 + 0.91 = 1.91 \quad \text{and} \quad I_{\text{tot}} = \frac{-5}{1.91} = -2.62\,\text{mA},$$

$$\text{so } I_{10} = -2.62\frac{1}{11} = -0.24\,\text{mA} \quad \text{and} \quad V_{10} = -2.4\,\text{V}$$

$$I_{10} = 0.48 + (-0.24) = 0.24 \quad \text{and} \quad V_{10} = 4.8 + (2.4) = 2.4\text{ V}$$

The exact values should be 0.25 mA and 2.5 V.

3. *Experiment.* Figure 3.30 shows the circuit and power sources.

Figure 3.31 shows the measured circuit and the voltage in the 10-kΩ resistor.

Its value is 2.34 V, as we expect.

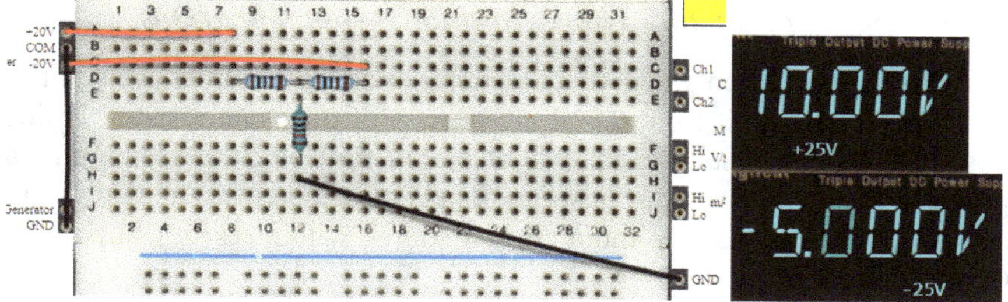

Figure 3.30. Circuit under experimentation for the superposition principle in VISIR.

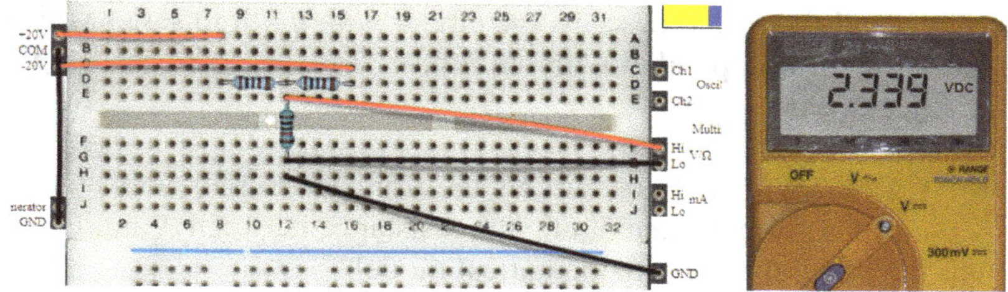

Figure 3.31. Voltage measurement in the circuit under experimentation for the superposition principle in VISIR.

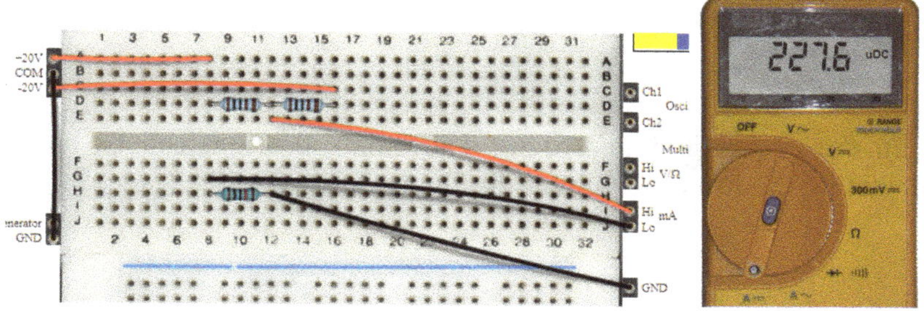

Figure 3.32. Current measurement in the circuit under experimentation for the superposition principle in VISIR.

Figure 3.32 shows the current measurement. As expected, we obtain 0.23 mA.

4. *Analysis and conclusion.* The experiment shows that the electric circuits (at least in this circuit) fulfill the superposition principle.

3.2.13 *DC Experiment: Thevenin and Norton Theorems*

1. *Introduction.* When we have an electric circuit, sometimes we want to "describe" the circuit from the point of view of only one resistor. Is this possible? In this case, the equivalent circuit will be the resistor plus a power source with a resistor in series (Thevenin theorem). The equivalent circuit will be the resistor and a current source with a resistor in parallel (Norton theorem).

Can we test these two theorems with VISIR?

2. *Mathematical model.* If we have a circuit with a load resistor (RL) between A and B, we can describe the circuit with a *V*th in series with the *R*th, or with an *I*th in parallel with the *R*th.

Mathematically,

- **Calculate *R*th**: Short-circuit the voltage sources and open the current sources, then obtain the resistance between A and B using the parallel–series mathematical model.
- ***V*th**: Open the current sources and calculate the V_{AB} using Ohm's and Kirchhoff's laws.
- ***I*th**: Calculate

$$I\text{th} = \frac{V\text{th}}{R\text{th}}$$

In the circuit in Figure 3.33, Thevenin's theorem can be applied. Calculating the *R*th of the circuit in Figure 3.34.

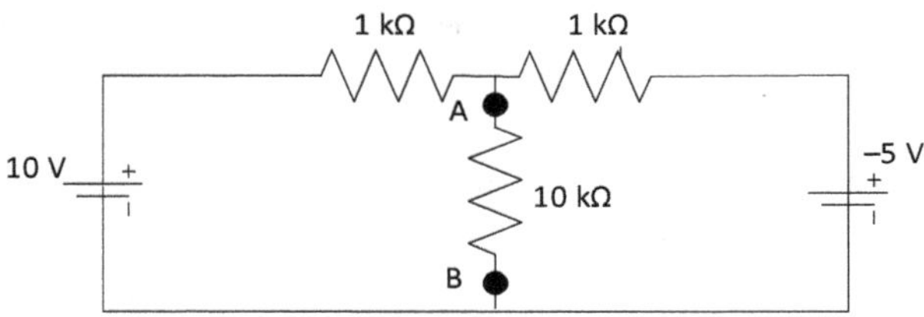

Figure 3.33. Circuit under experimentation for Thevenin's and Norton's theorems.

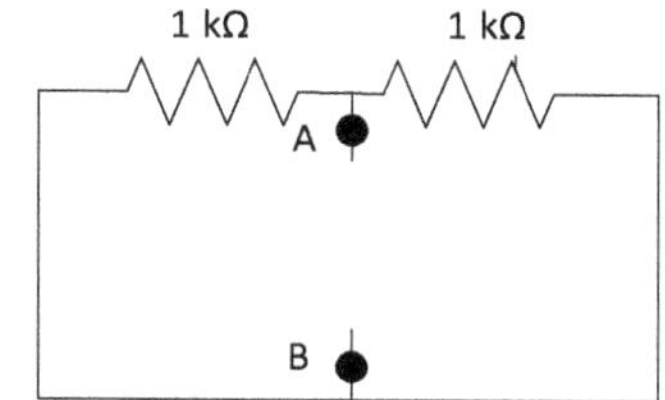

Figure 3.34. Thevenin's resistor.

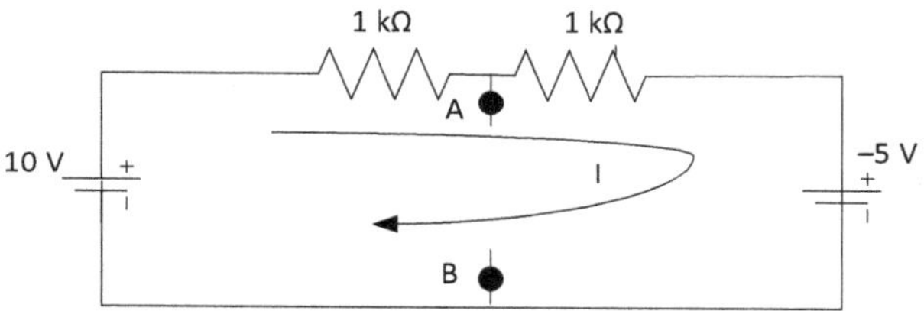

Figure 3.35. Thevenin's resistor calculation.

$$Rth \geq 1\ \text{k}\Omega \text{ in parallel with } 1\ \text{k}\Omega,\ \frac{1}{Rth} = \frac{1}{1} + \frac{1}{1},\ Rth = 0.5\ \text{k}\Omega$$

Calculate the *V*th and include it in Figure 3.35.

$$Vth \geq I = \frac{10-(-5)}{2} = 7.5\ \text{mA}, \text{ so } \quad Vth = V_{\text{AB}} = 10 - 7.5 \cdot 1$$
$$= 2.5\ \text{V} \text{ or } Vth = V_{\text{AB}} = 7.5 \cdot 1 + (-5) = 2.5\ \text{s}$$
$$Ith = \frac{V_{\text{AB}}}{Rth} = \frac{2.5}{0.5} = 5\ \text{mA}$$

The equivalent Thevenin's circuit is shown in Figure 3.36.

If the RL is 10 kΩ, then the voltage in RL is 2.38 and the current is 0.238 mA.

$$I = \frac{2.5}{10.5} = 0.238\,\text{mA}\ y V_L = 0.238 \cdot 10 = 2.38\,\text{V}$$

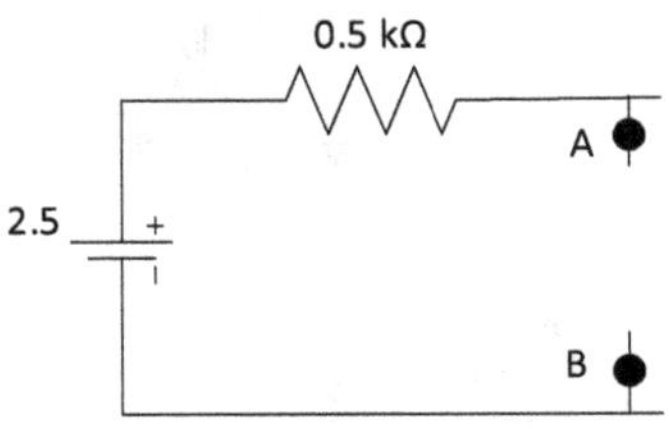

Figure 3.36. Thevenin's circuit.

3. *Experiment.* Figure 3.37 shows the drop voltage between A and B, the current between A and B, and the resistance between A and B.

If we include the 10-kΩ load resistor (Figure 3.38), then the voltage is as was established.

The current is around 0.25 mA (Figure 3.39), as was expected.

If we construct and measure the equivalent circuit, the results are as follows: 2.383 V and 0.234 mA, as was expected (Figure 3.40).

4. *Analysis and conclusion.* The experiment shows that Thevenin and Norton's theorems can be tested in the VISIR.

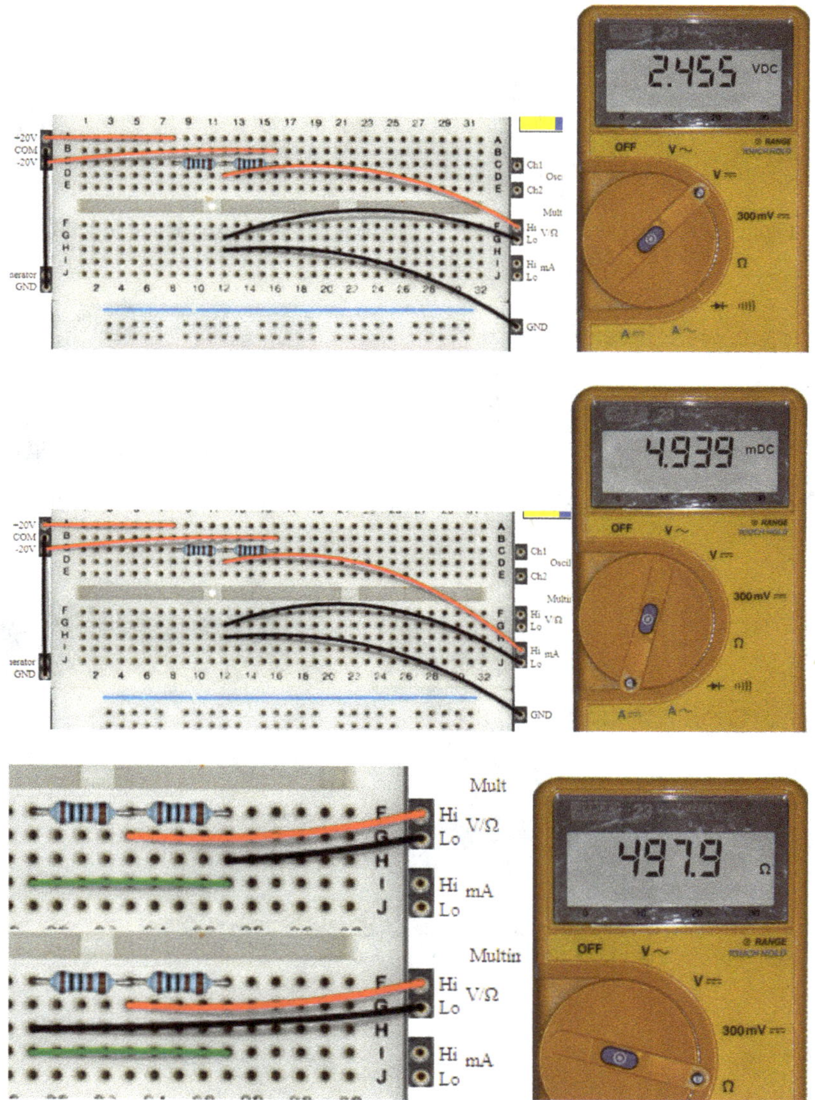

Figure 3.37. Thevenin's circuit experiment in VISIR.

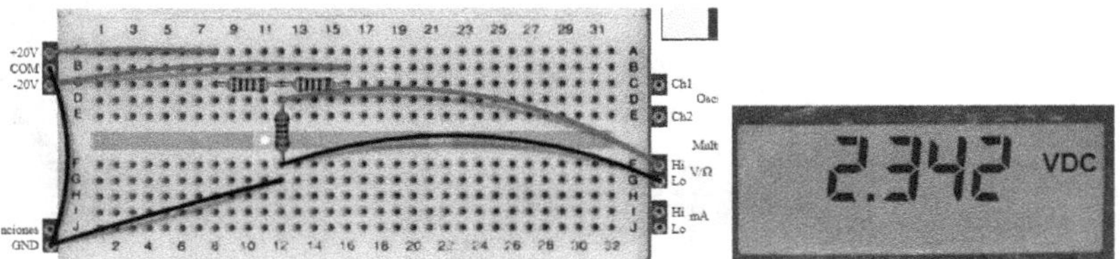

Figure 3.38. Voltage measurement in Thevenin's circuit experiment in VISIR.

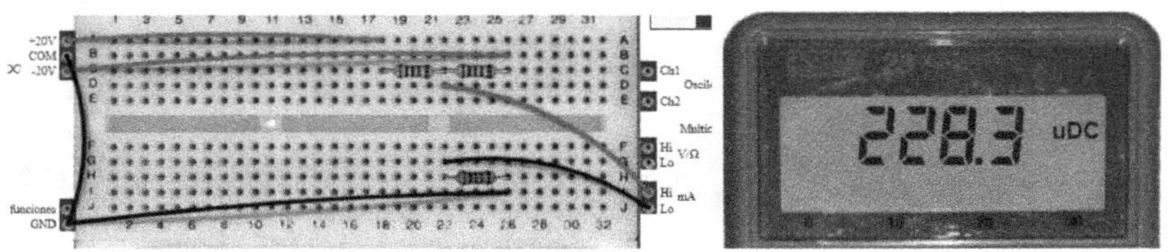

Figure 3.39. Current measurement in Thevenin's circuit experiment in VISIR.

Figure 3.40. Measurements in Thevenin's equivalent circuit.

3.3 AC Circuits

The set of experiments and circuits is as follows:

- AC Experiment: Measuring AC signals
- AC Experiment: Maximum power transfer theorem
- AC Experiment: Ohm's law and Kirchhoff's laws
- AC Experiment: Capacitor, capacitance, and reactance
- AC Practice: Measuring an RC circuit
- AC Practice: RC circuit as a low-pass filter
- AC Experiment: Cut-off frequency and time constant of an RC circuit
- AC Practice: Using the cut-off frequency and time constant to design circuits
- AC Experiment: Characterization of RL circuits.

3.3.1 *AC Experiment: Measuring AC Signals*

1. *Introduction.* An AC signal varies over time. The most popular AC signal in circuits is the sinusoidal signal because it consists of a single frequency. The two main values of a sinusoidal wave are the frequency (or period) and the peak-to-peak voltage (V_{pp}) or the maximum voltage (V_{max}), assuming there is no superimposed DC voltage. This value is an instant value; it is the maximum voltage in the time series. V_{pp} is defined as $V_{max} - V_{min}$; if $[V_{max}]$ is equal to $[V_{min}]$, then $V_{pp} = 2 \cdot V_{max}$.

In general, the average value is better than the maximum value because an area (the average) is more representative than an instant value. However, this is not always true, for instance, with sinusoidal waves.

Two different AC signals can be different (e.g., triangular and sinusoidal waves) and have the same maximum value, so the maximum value is not a good descriptor of the signal.

At the same time, however, the average is always 0 because the AC signal is an alternated signal: half period is positive, and half period is negative. The concept of average is correct (the area) but the calculus is not. In this case, we should obtain the V_{rms}, root main square voltage. The rms voltage is also known as the quadratic mean.

Using VISIR, we can generate and visualize/measure AC signals.

2. *Mathematical model.* The rms voltage is the arithmetic mean of the squares of a set of voltages in a period. We can calculate V_{rms} using the following (generic) equation:

$$V_{ef} = \sqrt{\frac{\int_0^T v(t)^2}{T}} = \sqrt{\frac{\int_0^T (\text{A} \cdot \text{sen}\,(\omega \cdot t))^2}{T}} = \frac{A}{\sqrt{2}} = \frac{V_{max}}{\sqrt{2}}$$

3. *Experiment.* The experiment is very simple: we generate a sine wave with a period and peak-to-peak value, we measure V_{rms}, and finally we check whether the previous equation is correct.

Using an inquiry approach, the students do not see the previous equation; they generate and measure different sine waves and finally try to discover the mathematical relation between V_{max} and V_{rms}.

In the first experiment, we maintain the period and change V_{pp} to observe and measure the sinusoidal wave. In the following experiment in Figure 3.41, we see the function generator (FG) and the oscilloscope.

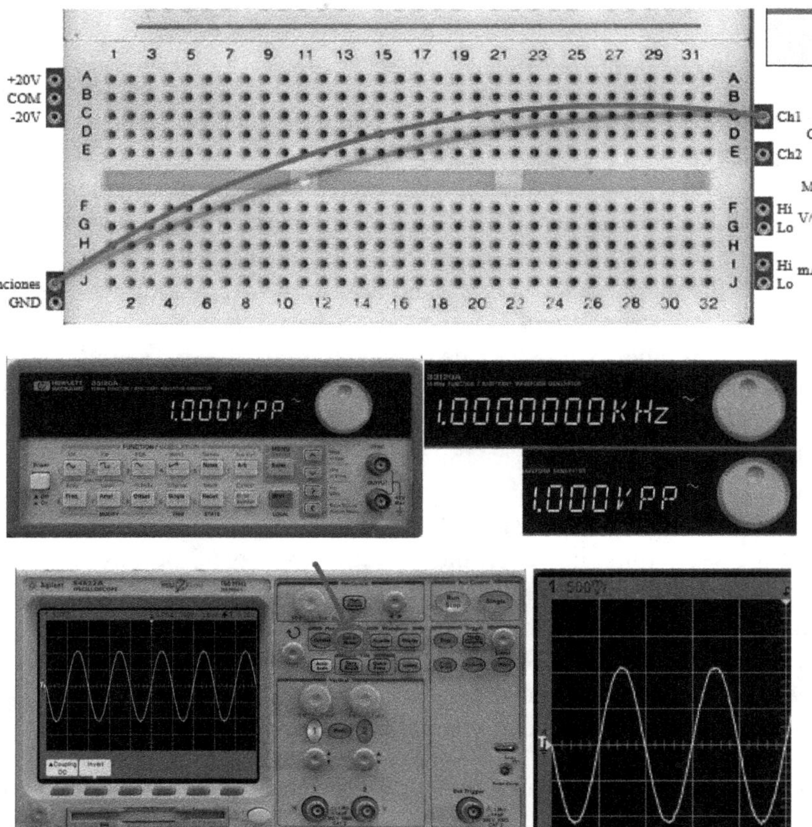

Figure 3.41. Function generator and oscilloscope in VISIR.

Before continuing with the experiment, it is important to note that if we look carefully at the oscilloscope, we see that the signal is double the expected value. Each square is 0.5 V, so the maximum is 1 V, and the minimum is –1 V, then V_{pp} = 2 V. It is exactly double 2 V. This will be explained in Experiment 3.3.2, but for the moment, it is important to learn how to use the VISIR FG: If you want V_{pp} = 2 V, then choose 1 V; instead of reading V_{pp}, read V_{max} in the FG.

To measure V_{max}, V_{pp}, and V_{rms}, click on Quick Meas., click on Select as many times as needed to reach the desired value (click on the gray button, not in the text), and click on Measure. Do this three times: Maximum, peak–peak, and rms. Remember to click on Perform Experiment (Figure 3.42).

For V_{pp} = 2 V and V_{pp} = 5 V, the results are shown in Figure 3.43.

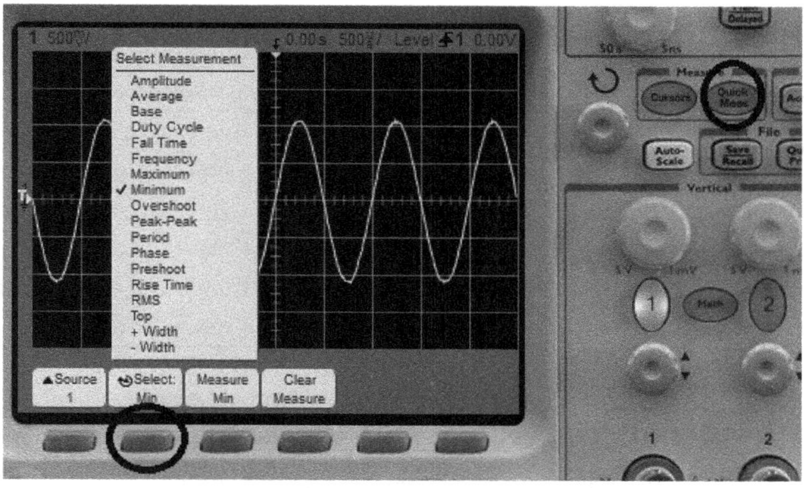

Figure 3.42. Measurements in the oscilloscope in VISIR.

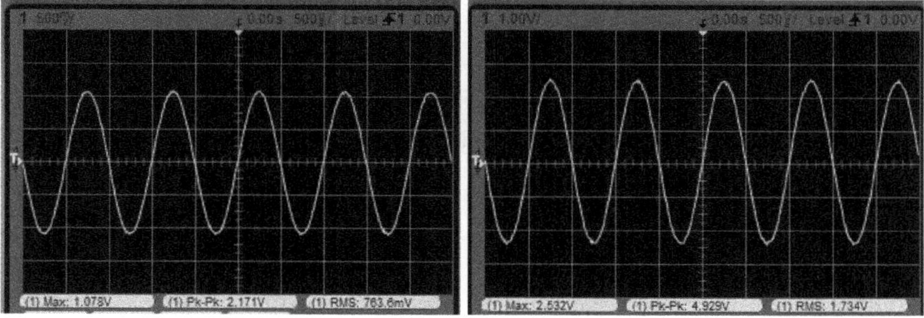

Figure 3.43. Waveforms in the oscilloscope of VISIR.

In Table 3.10, you can see the results for different V_{pp}. The values in the last three rows are measurements.

The first rms row is the measured value, and the last rms row is the calculated value with $V_{max}/1.4142$ (V_{max} is the theoretical value). These two rows are similar. The value in the Relation row is the quotient between the measurements Vmax and Vrms. This Relation value is around $\sqrt{2}$, and this is correct.

We can also see that the measured Vmax differs from the expected V_{max}. For instance, in the third column, instead of 1.5 V, VISIR measures 1.599. This is real, and remember that V_{max} is the maximum value reached by the sinusoidal wave during the time shown in the graphic; it is not the mean, it is the maximum.

It is important to note that the oscilloscope obtains the measurements by looking at the graphic. That is, if the scale is not correct, the measurement will not be correct. For instance, Figure 3.44 shows the VISIR

Table 3.10. Relation between V_{pp} and V_{rms} for sinusoidal waves.

V_{pp}	1 V	2 V	3 V	4 V	5 V	8 V	10 V
V_{max}	0.520	1.078	1.599	2.093	2.532	4.052	5.088
V_{pp}	1.034	2.171	3.248	4.051	4.929	7.934	10.17
V_{rms}	0.362	0.764	1.140	1.399	1.734	2.773	3.569
Relation	1.436	1.411	1.402	1.496	1.460	1.461	1.426
V_{rms}	0.354	0.707	1.061	1.414	1.768	2.828	3.535

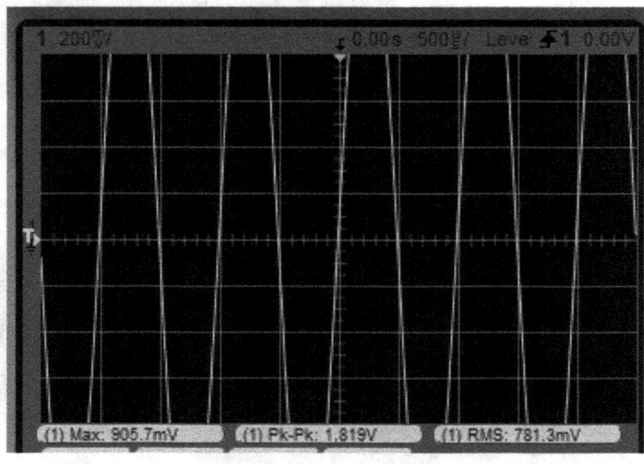

Figure 3.44. Incorrect use of the VISIR oscilloscope.

oscilloscope for a 3-V_{pp} sinusoidal wave, then V_{max} should be around 1.5 V and V_{rms} around 1.061 V. However, as the scale is 200 mV, V_{max} is 0.905 V, and V_{rms} is 0.781 V. If the scale were 0.5 V, then the results would be correct.

In the second experiment, we change the frequency (i.e. the period) and we maintain V_{pp} as 10 V. Table 3.11 shows different results. The expected value is 3.535 V, and all the obtained measurements are around this value.

4. *Analysis and conclusion.* Any AC wave can be characterized using its instant values: V_{max}/V_{pp} and its V_{rms}. V_{rms} is the value to characterize an AC signal.

In a sine wave, V_{rms} can be obtained as $V_{rms} = \frac{V_{max}}{\sqrt{2}} = \frac{V_{pp}}{2 \cdot \sqrt{2}}$. This expression is valid for any V_{max} and for any frequency/period.

Similarly to the experiment with a sine wave, we can measure V_{rms} of a triangular/sawtooth/square wave. Instead of a sinusoidal wave, we choose triangular/square/sawtooth (Figure 3.45).

The question is as follows: Can we find the mathematical expressions to obtain V_{rms} of each wave using V_{max}/V_{pp}? The way is to systematically repeat the previous experiment using a different way.

Table 3.11. Relation between V_{rms} and frequency.

f (Hz)	50	100	200	500	1000	2000
V_{rms} (V)	3.578	3.590	3.584	3.572	3.568	3.569

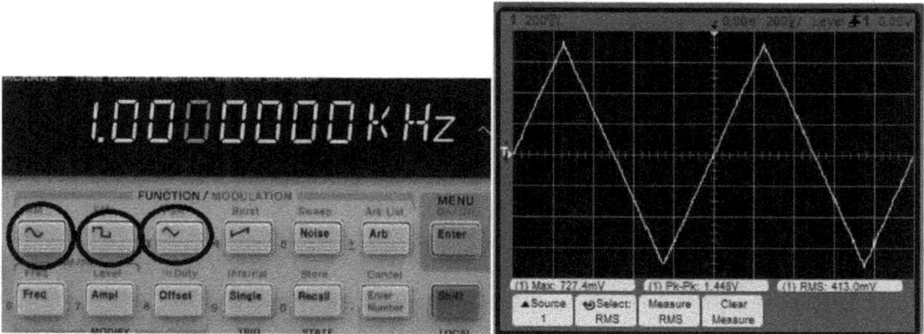

Figure 3.45. Triangular waveform.

3.3.2 *AC Experiment: Maximum Power Transfer Theorem*

1. *Introduction.* In the previous experiment, we have seen that the function generator gives us double the "expected" V_{pp}. This situation can be explained using the maximum power transfer theorem.

With this theorem, we understood that the maximum power is obtained when the output resistance is equal to the internal resistance of the DC power supply.

2. *Mathematical model.* The output voltage of the following circuit is

$$I_{rms} = \frac{V_{ACrms}}{R_{tot}} = \frac{V_{ACrms}}{R_s + R_o}$$

$$V_{orms} = I_{rms} \cdot R_o = V_{ACrms} \cdot \frac{R_o}{R_s + R_o} = V_{ACrms} \cdot \frac{R_o}{50 + R_o}$$

V_o is a part of the input voltage, and this part depends on the relation between R_s and R_o. Table 3.12 shows the calculated values for different R_o.

3. *Experiment.* The experiment in Figure 3.46 is very simple: Change the output resistor in the circuit and measure the rms value of the output voltage. The input is 3 V_{pp}; this value is the one selected in the function generator.

Table 3.13 shows the expected rms voltages for different R_o.

The first graphic in Figure 3.47 is with $R_o = 1000\ \Omega$, and we can see that all the input voltage drops in the output because R_o is much higher than the R_s ($1000\ \Omega > 50\ \Omega$). In the second graphic, however, $R_o = 6\ \Omega$, so the output voltage is low, around 0.23 V in rms.

Table 3.12. V_o for different R_o.

R_o (Ω)	6	50	100	0.5	1	10
V_o (V) Calculated	0.227	1.061	1.414	1.928	2.02	2.111

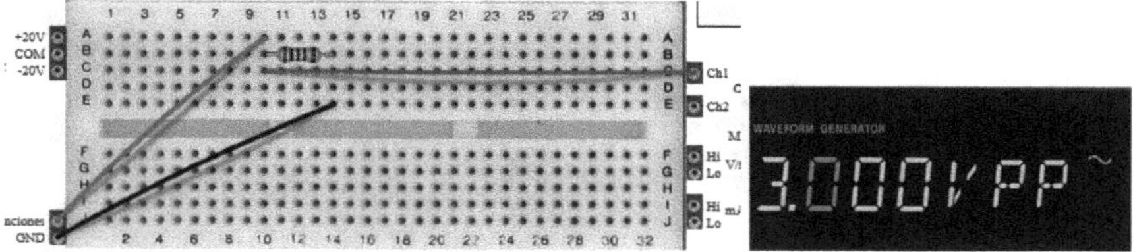

Figure 3.46. Circuit for the maximum power transfer theorem experiment.

Table 3.13. Calculated and measured voltages.

R_o (Ω)	6	50	100	0.5	1	10
V_o (V) Measured	0.269	1.026	1.384	1.887	1.981	2.075
V_o (V) Calculated	0.227	1.061	1.414	1.928	2.02	2.111

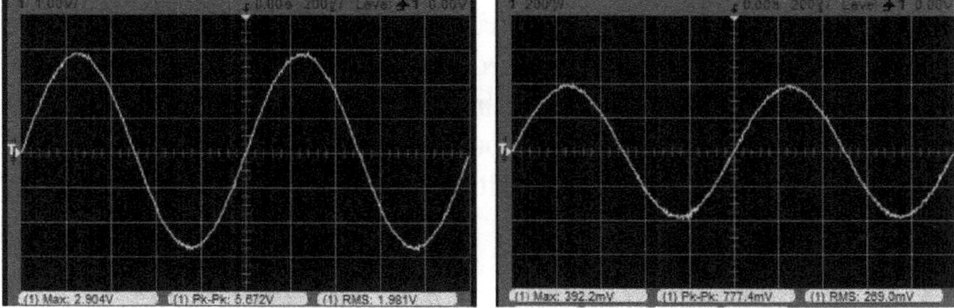

Figure 3.47. Waveforms for the maximum power transfer theorem experiment.

We see that only with R_o = 50 Ω, the output voltage is equal to the selected one in the function generator, i.e., only when $R_o = R_s$. In Figure 3.48, with R_o = 50 Ω (two resistors of 50 Ω in parallel), we can see that output has 3 V_{pp}, as we expected at the beginning, and not 6 V_{pp}.

4. *Analysis and conclusion.* A DC power supply or a function generator powers circuits, but the input voltage depends not only on the selected value in the power source but also on the characteristics of each circuit.

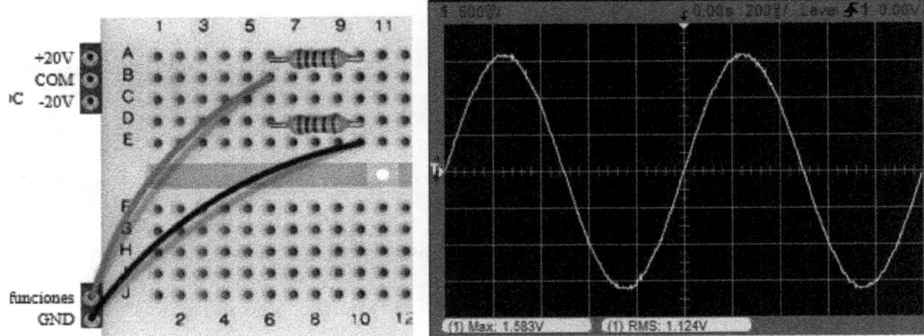

Figure 3.48. Waveform for the maximum power transfer theorem experiment.

There is an important recommendation: After powering the circuit, measure the voltage to see if its value is as expected.

Some DC power sources and function generators have a knob to select the amplitude, but in the instrument, there is no number/digit. The manufacturer avoids "saying" that you have 3 V, because this may not be true, or maybe the manufacturer should say something like "the input voltage selected is only true if the output resistor is 50 Ω, otherwise you need to measure or calculate the voltage value".

3.3.3 *AC Experiment: Ohm's Law and Kirchhoff's Laws*

1. *Introduction.* Ohm's and Kirchhoff's laws were established with DC, so what happens if the current is alternating?

2. *Mathematical model.* Ohm's and Kirchhoff's laws were presented in previous experiments.

$$V = I \cdot R, \quad I = \frac{V}{R}, \quad R = \frac{V}{I}$$
$$I_{\text{tot}} = I_{R1} + I_{R2} + I_{R3} + \cdots + I_{Rn}$$
$$V_{\text{DC}} = V_{R1} + V_{R2} + V_{R3} + \cdots + V_{Rn}$$

3. *Experiment.* We will simply work with two circuits with 1 kΩ and 10 kΩ, series and parallel, and we will measure currents and voltage drops in

both circuits (Figure 3.49). To measure the signals we will use the multimeter, not the oscilloscope.

All the measurements in Table 3.14 are rms (root mean square) values. In the multimeter in Figure 3.50, we need to choose the alternating symbol (V~). The two circuits are powered with a 3 V_{pp} and 1 kHz sine wave (with no DC offset).

4. *Analysis and conclusion.* The previous experiment confirms that Ohm's and Kirchhoff's laws are also valid for alternating input signals.

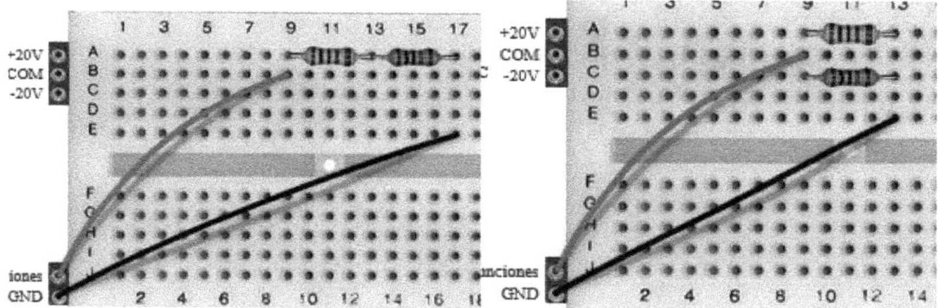

Figure 3.49. Ohm's and Kirchhoff's laws for AC circuits.

Table 3.14. Voltage measurements in two AC circuits.

Series Circuit				Parallel Circuit				
V_i	I_{tot}	V_{1k}	V_{10k}	V_i	I_{tot}	I_{1k}	I_{10k}	$V_{1k/10k}$
2.111	0.21	0.193	1.916	2.008	2.235	2.026	0.202	2.008

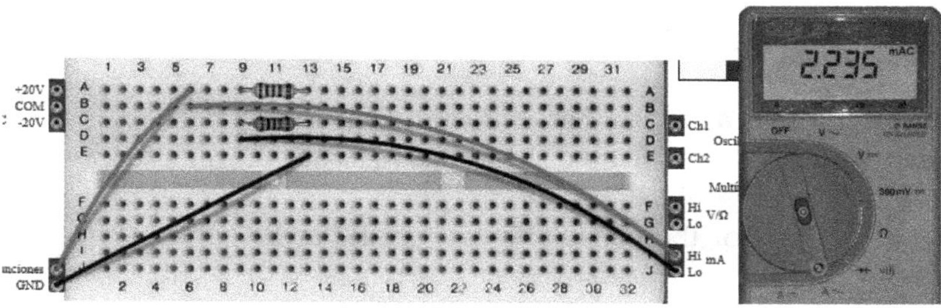

Figure 3.50. Ohm's and Kirchhoff's laws for AC circuits: Current measurement.

3.3.4 *AC Experiment: Capacitor, Capacitance, and Reactance*

1. *Introduction.* In AC circuits, we can not only find resistors but also capacitors, inductors, diodes, etc.

A capacitor has multiple utilities, so the first step is to characterize it. A capacitor has a capacitance (*C*), which is measured in farads (F). However, the reactance, XC, of a capacitor depends not only on *C* but also on the frequency of the AC signal. The objective of the experiment is to measure this behavior.

2. *Mathematical model.* The capacitive reactance is measured in Ω and can be calculated with the following equation:

$$X_{\mathrm{C}} = \frac{1}{j\omega C} = \frac{-j}{\omega C}, \qquad \text{where } \omega = 2\pi f$$

3. *Experiment.* The experiment is very simple: we set an RC circuit (resistor and capacitor), we power it, we measure V_{C} and I_{C} in rms, and we obtain X_{C} using Ohm's law

$$X_{\mathrm{C}} = \frac{V_{\mathrm{C}}}{I}, \quad \text{this is the module } |X_{\mathrm{C}}|$$

The RC circuit in Figure 3.51 has a 1-kΩ resistor and a 1-μF capacitor and is powered with a sine wave of 3 V_{pp} and different frequencies.

When we have the circuit shown in Figure 3.52, we have to change the values of the capacitor and of the frequency. Table 3.15 shows the results for two capacitor values and three frequencies. The calculated value is written in *italics*.

4. *Analysis and conclusion.* The experiment shows that the behavior of the capacitor depends on *C* (capacitance) and *f* (frequency). This is important for the following applications.

The module of the capacitive reactance can be calculated with the previous formula.

Figure 3.51. Experiment for measuring the capacitance.

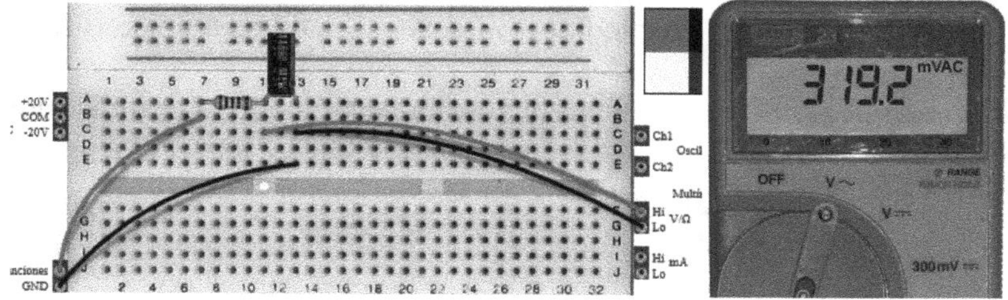

Figure 3.52. Voltage measurement in the capacitance experiment.

Table 3.15. Capacitances for different frequencies.

	100 Hz			200 Hz			500 Hz		
	V_C (V)	I (mA)	X_C (Ω)	V_C (V)	I (mA)	X_C (Ω)	V_C (V)	I (mA)	X_C (Ω)
1 μF	1.745	1.119	1559	1.257	1.599	786.1	0.615	1.913	321.5
			1592			*795.8*			*318.3*
10 μF	0.319	2.004	159.2	0.163	2.024	80.53	0.671	2.031	33.04
			159.2			*79.58*			*31.83*

Other additional experiments can be performed:

- What happens with the resistance? Does it depend on frequency or not?
- What happens if we change the wave type to triangular/square/ sawtooth?
- What happens to the resistor? If we change the resistor, does X_C change?

3.3.5 *AC Practice: Measuring an RC Circuit*

1. *Introduction.* We can do the same with RC circuits as in the experiment with DC circuits. In this case, the circuit includes one resistor in series with one capacitor.

This circuit is very simple but is also very useful as a noise filter.

2. *Mathematical model.* The analytical solution of the RC circuit is shown in Figure 3.53.

$$Z = R + X_C \cdot j = M\angle\theta$$

$$\text{where } M = \sqrt{\frac{(R\cdot\omega\cdot C)^2+1}{(\omega\cdot C)^2}} \quad \text{and} \quad \theta = \text{arctg}\left(-\frac{1}{\omega\cdot C\cdot R}\right)$$

$$i(t) = \frac{v(t)}{Z_{tot}} = \frac{V_{max}\cdot \text{sen}(\omega\cdot t)}{\sqrt{\frac{(R\cdot\omega\cdot C)^2+1}{(\omega\cdot C)^2}}\angle\text{arctg}\left(-\frac{1}{\omega\cdot C\cdot R}\right)}$$

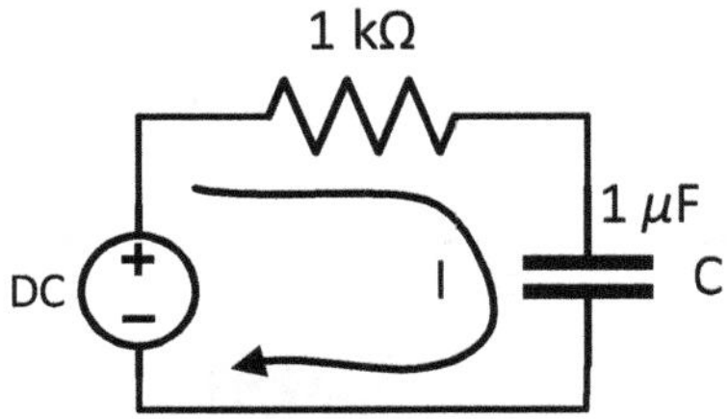

Figure 3.53. Basic RC circuit.

$$v_R(t) = \frac{R \cdot \omega \cdot C}{\sqrt{(R \cdot \omega \cdot C)^2 + 1}} \cdot V_{max} \cdot \text{sen}\left(\omega \cdot t - \text{arctg}\left(-\frac{1}{\omega \cdot C \cdot R}\right)\right)$$

$$v_C(t) = \frac{1}{\sqrt{(R \cdot \omega \cdot C)^2 + 1}} \cdot V_{max} \cdot \text{sen}\left(\omega \cdot t - \text{arctg}\left(-\frac{1}{\omega \cdot C \cdot R}\right) - \frac{\pi}{2}\right)$$

If $C = 1\ \mu$F, $= 1$ kΩ, $f = 200$ Hz, $V_{max} = 5$ V

$$v_i(t) = 5 \cdot \sin 2 \cdot \pi \cdot 200 \cdot t$$

$$Z = 1000 - 796j = 1277\angle -0.67 \text{ rd } \Omega$$
$$i(t) = 3.91 \cdot \text{sen}(2 \cdot \pi \cdot 200 \cdot t + 0.67) \text{ mA}$$
$$v_R(t) = 3.91 \cdot \text{sen}(2 \cdot \pi \cdot 200 \cdot t + 0.67) \text{ V}$$
$$v_C(t) = 3.11 \cdot \text{sen}(2 \cdot \pi \cdot 200 \cdot t + 0.90) \text{ V}$$

3. *Practice*. We have to design, power, and measure the circuit (Figure 3.54).

In Figure 3.55, we can see the two signals: input (Channel 1) and output (Channel 2). The input signal is sinusoidal, as is the output. Sometimes it is better to look at only one of the signals, by simply clicking on 1 or 2. To measure the values in the output, we must change the Source to 2.

Now we have to measure values one by one (Table 3.16).

We have completed only two values in Table 3.16. For recording the current values, we need to use the multimeter because the oscilloscope does not measure current (Figure 3.56).

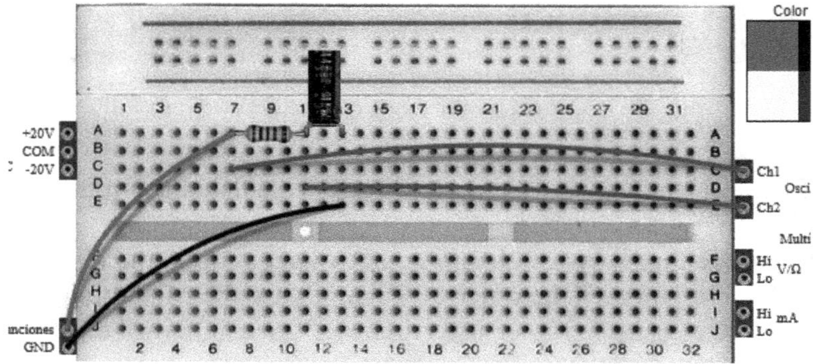

Figure 3.54. RC circuit in VISIR.

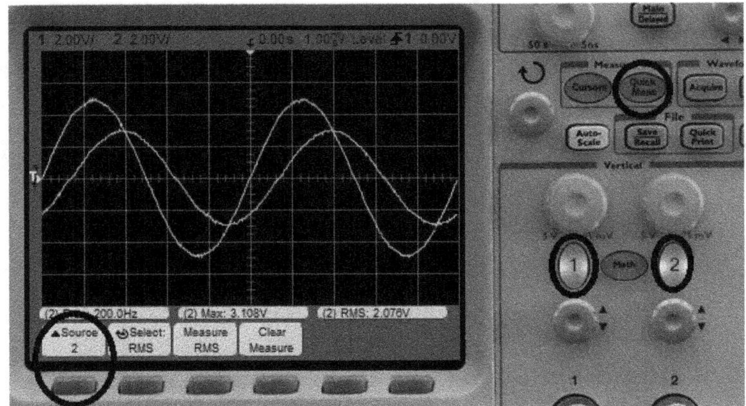

Figure 3.55. RC circuit in VISIR oscilloscope.

Table 3.16. Measurements for the RC circuit I.

Circuit						Capacitor			Resistor	
f (Hz)	Z_{tot} (Ω)	Z_{tot} (rd)	X_C (Ω)	I_{rms} (mA)	I_{max} (mA)	V_{rms} (V)	V_{max} (V)	V_c (rd)	V_{rms} (V)	V_{max} (V)
200						2.076	3.108			

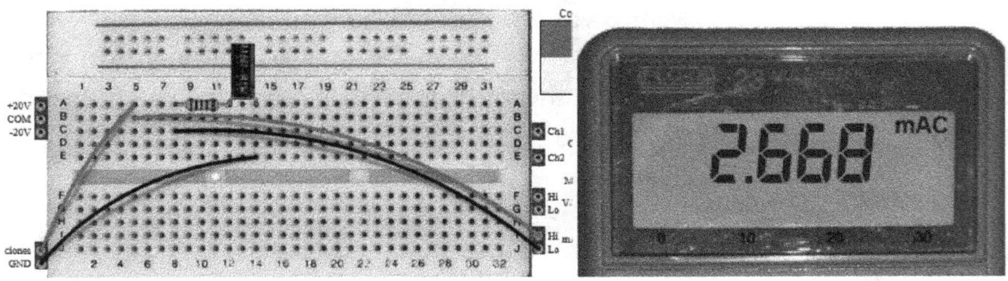

Figure 3.56. Current measurement of the RC circuit in VISIR.

Table 3.17 now includes I_{rms}, and I_{max} is calculated using the expression $I_{max} = I_{rms} \cdot \sqrt{2}$.

There are four ways of obtaining the voltage drop in the resistor: use the multimeter, modify the order of R and C, use the Math option in the oscilloscope (Figure 3.57), or use Kirchhoff's second law, i.e., $V_i = V_r + V_c$ (Table 3.18).

To obtain the X_C and Z_{tot} values, we apply Ohm's law (Table 3.19).

Table 3.17. Measurements for RC circuit II.

Circuit						Capacitor			Resistor	
f (Hz)	Z_{tot} (Ω)	Z_{tot} (rd)	X_C (Ω)	I_{rms} (mA)	I_{max} (mA)	V_{rms} (V)	V_{max} (V)	V_c (rd)	V_{rms} (V)	V_{max} (V)
200				2.668	3.773	2.076	3.108			

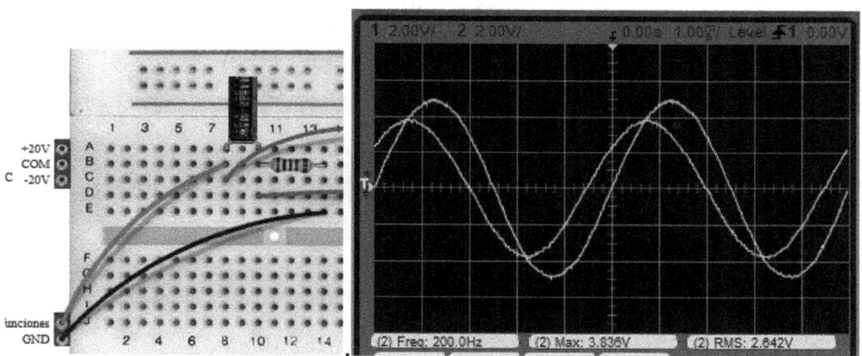

Figure 3.57. Output voltage measurement of the RC circuit in VISIR.

Table 3.18. Measurements for the RC circuit III.

Circuit						Capacitor			Resistor	
f (Hz)	Z_{tot} (Ω)	Z_{tot} (rd)	X_C (Ω)	I_{rms} (mA)	I_{max} (mA)	V_{rms} (V)	V_{max} (V)	V_c (rd)	V_{rms} (V)	V_{max} (V)
200				2.668	3.773	2.076	3.108		2.642	3.836

Table 3.19. Measurements for the RC circuit IV.

Circuit						Capacitor			Resistor	
f (Hz)	Z_{tot} (Ω)	Z_{tot} (rd)	X_C (Ω)	I_{rms} (mA)	I_{max} (mA)	V_{rms} (V)	V_{max} (V)	V_c (rd)	V_{rms} (V)	V_{max} (V)
200			778	2.668	3.773	2.076	3.108		2.642	3.836

Table 3.20. Measurements for the RC circuit V.

Circuit						Capacitor			Resistor	
f (Hz)	Z_{tot} (Ω)	Z_{tot} (rd)	X_C (Ω)	I_{rms} (mA)	I_{max} (mA)	V_{rms} (V)	V_{max} (V)	V_c (rd)	V_{rms} (V)	V_{max} (V)
200	1330	0.69	778	2.668	3.773	2.076	3.108		2.642	3.836

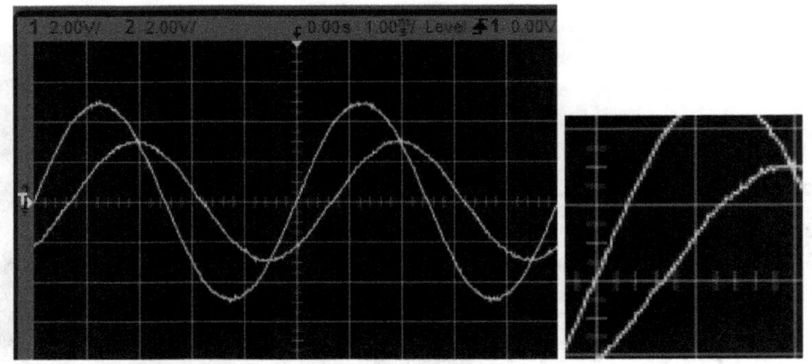

Figure 3.58. Delay measurement in the RC circuit.

$$X_C = \frac{2.076}{2.668} = 778\Omega \quad \text{and} \quad Z_{tot} = \frac{5}{3.773} = 1.33\ \text{k}\Omega$$

Finally, we need to obtain the output delay (Table 3.20). Looking in detail at the output signal, we can measure the delay (Figure 3.58). On the left-hand side, we see that the period is 5 squares and each square is 1 ms (T = 5 ms, so f = 200 Hz), and on the right, we see that the output is delayed by more or less 3.5 squares, so 0.7 ms. We can convert this time delay into radians $\arg(V_c) = \frac{0.7\cdot 2\cdot \pi}{5} = 0.88\,\text{rd}$, so $\arg(Z_{tot}) = \frac{\pi}{2} - 0.88 = 0.69\,\text{rd}$.

The circuit, after the measurements, can be analytically described as follows:

$$v_i(t) = 5 \cdot \sin 2 \cdot \pi \cdot 200 \cdot t, \quad R = 1\text{k}\ y\ C = 1\ \mu\text{F}$$
$$|X_C| = 778\ \Omega,\ \arg(X_C) = -1.57\ \text{rd}$$
$$|Z_{tot}| = 1330\ \Omega,\ \arg(X_C) = -0.69\ \text{rd}$$
$$i(t) = 3.773 \cdot \sin(2 \cdot \pi \cdot 200 \cdot t + 0.69)\ \text{mA}$$
$$v_r(t) = 3.773 \cdot \sin(2 \cdot \pi \cdot 200 \cdot t + 0.69)\ \text{V}$$
$$v_c(t) = 3.108 \cdot \sin(2 \cdot \pi \cdot 200 \cdot t - 0.88)\ \text{V}$$

As can be tested, the expressions obtained using VISIR are similar to the expressions obtained analytically.

4. *Analysis and conclusion.* Using VISIR and the basic laws, we can obtain the mathematical expressions of an RC circuit.

We have seen that in an RC circuit, if the input is a sinusoid, then the output will be a sinusoid with the same frequency, lower V_{max}, and delayed. In the next practice, we will use this conclusion.

The previous practice was with a sinusoidal wave input. What happens if the input is a triangular wave? Will the output be a triangular wave with the same frequency or smaller? What happens now? Why?

3.3.6 *AC Practice: RC Circuit as a Low-pass Filter*

1. *Introduction.* As seen in the previous practice, in an RC circuit, if the input is a sinusoid, then the output will be a sinusoid of the same frequency, smaller, and with a delay.

Consider AC in the plug: 220 V_{rms} and 50 Hz (in Europe). What happens if there is noise? Noise in general is a high-frequency signal that is added to the original.

The question is as follows: Can we filter the noise signal without affecting the original signal? The answer is yes, using a simple RC circuit.

2. *Mathematical model.* As we recall, the reactance is affected by the frequency: The higher the frequency, the lower the reactance; and the resistance is unaffected by the frequency.

$$X_C = \frac{-j}{2 \cdot \pi \cdot f \cdot C} \quad X_R = R$$

Thus, if we have two sinusoidal inputs — $v_i(t)$ and $v_n(t)$ — with different frequencies — f_i and $f_n f_n >> f_i$ — then, according to the superposition principle, the output will be the sum of the outputs for each input: the plug and the noise.

$$v_c(t) = v_{cfi}(\mathrm{t}) + v_{c\,fn}(t)$$

Now what happens with XC at different frequencies? Looking at the previous formula, we see that the high-frequency signals are rejected because the value of XC decreases compared to R. And vice versa: If the signal frequency is low, XC increases, and the output voltage increases. In short, a high-frequency signal is filtered, i.e., it is rejected.

To measure the effect of the filter, we use the following formula:

$$\%\text{output} = \frac{V_{\text{rms,output}}}{V_{\text{rms,input}}} = \frac{V_{\text{rms,C}}}{V_{\text{rms,i}}} \quad \text{and} \quad \% = 100 - \%\text{output}$$

These values can be easily measured in VISIR.

3. *Practice.* The practice is very simple: An RC circuit (Figure 3.59) is excited with sinusoidal signals of different frequencies and we measure the rms voltages at the input and at the output, in the capacitor.

The RC circuit is made with $R = 1\ \text{k}\Omega$ and $C = 1\ \mu\text{F}$; the sinusoidal inputs will be 10 V_{pp} of different frequencies: 100 Hz, 200 Hz, 500 Hz, and 1000 Hz. At 500 Hz, we obtain Table 3.21 with the data in Figure 3.60.

Looking at Table 3.21, we see that the output attenuation (effect of the filter) is higher as the frequency increases.

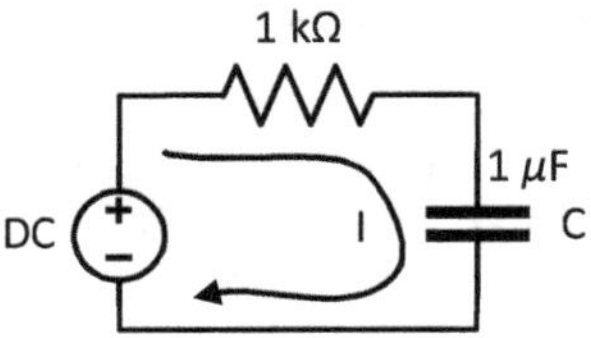

Figure 3.59. RC circuit as a low-pass filter.

Table 3.21. Low-pass filter data and measurements.

f (Hz)	100	200	500	1000
$V_{i,rms}$ (V)	3.53	3.47	3.42	3.40
$V_{o,rms}$ (V)	2.92	2.08	1.00	0.51
% output	82	60	29	15
% filter	18	40	71	85

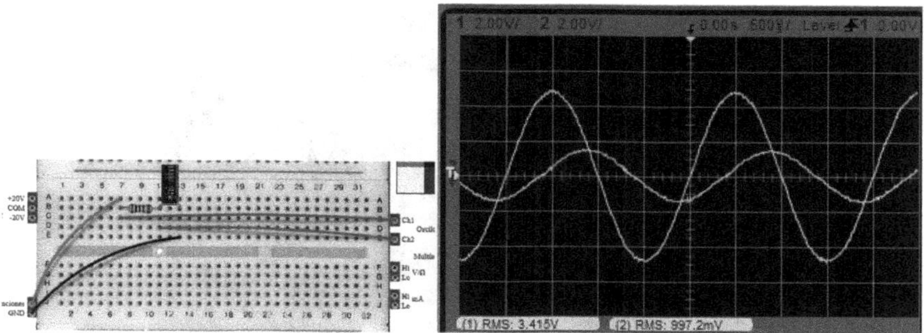

Figure 3.60. RC circuit as a low-pass filter: Waveform.

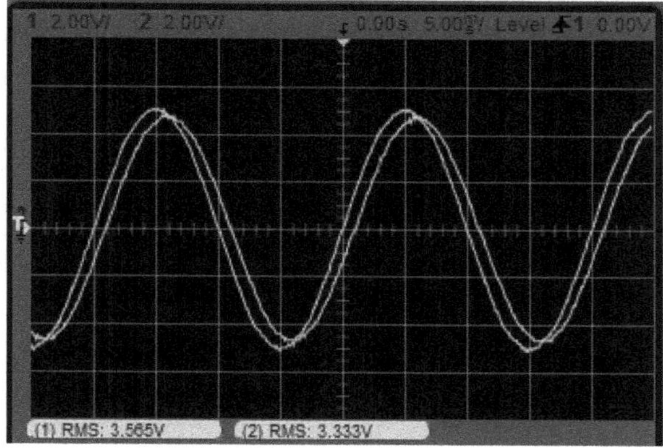

Figure 3.61. RC circuit as a low-pass filter: 50 Hz input.

Now we can imagine that we have in the plug a 10 V_{pp} (instead of 220 V_{rms}) at 50 Hz and there is a 1000-Hz and 1-V_{pp} noise. With VISIR, we cannot generate two sinusoidal signals at the same time; we need to experiment with one sinusoidal signal at a time.

With $R = 1$ kΩ and $C = 1$ μF, the input is 10 V_{pp} and 50 Hz. In Figure 3.61, we can see that the output is almost the input, the % output = $\frac{3.333}{3.566} = 93\%$. Thus, only 7% of the input has been lost. $v_{c\ \text{input}} = 4.71 \cdot \sin(2 \cdot \pi \cdot 50 \cdot t - \text{delay})$.

However, if we excite the RC circuit with a 1-V_{pp} and 1000-Hz sinusoid, we obtain Figure 3.62. We can see that the input is almost totally

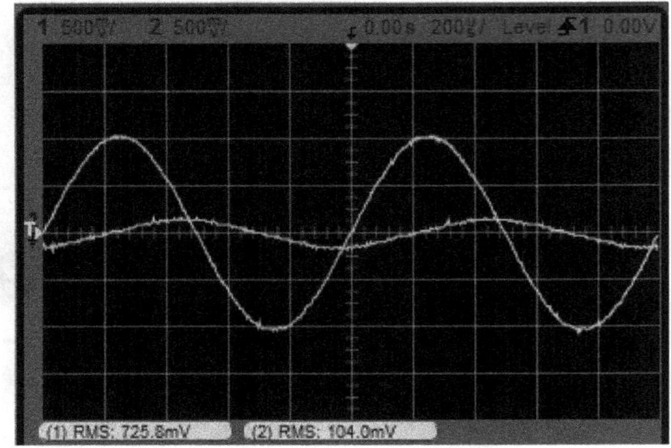

Figure 3.62. RC circuit as a low-pass filter: 1000 Hz input.

rejected; the output is % output $= \frac{0.104}{0.753} = 14\%$. Thus, 86% of the noise has been filtered. $v_{c\ \text{input}} = 0.15 \cdot \sin(2 \cdot \pi \cdot 1000 \cdot t - \text{delay})$.

Therefore, the RC circuit output will be

$$v_0(t) = 4.71 \cdot \sin(2 \cdot \pi \cdot 50 \cdot t - \text{delay}) + 0.15 \cdot \sin(2 \cdot \pi \cdot 1000 \cdot t - \text{delay})$$

Noise has been more or less filtered, and the input has also been more or less maintained. In this practice, obtaining the delay is not important.

4. *Analysis and conclusion.* We have seen that a simple RC circuit can be used as an effective high-frequency filter. The RC circuit is a noise filter.

Additionally, we recommend analysis of the behavior of a CR circuit. Instead of R + C in series, the circuit will be C + R in series.

3.3.7 *AC Experiment: Cut-off Frequency and Time Constant of an RC Circuit*

1. *Introduction.* The cut-off frequency and the time constant are two values that describe the behavior of an RC circuit. These two values are used in the process design.

2. *Mathematical model.* If the output of an RC circuit is 70.7% of the input, then the frequency of the input is called the cut-off frequency, f_c. This value can be obtained with the following formula:

$$f_c = \frac{1}{2 \cdot \pi \cdot R \cdot C}$$

If the input of an RC circuit is a square wave, then the output will be a saturated exponential. The instant the output reaches 63% of the final value is the time constant, τ. It can be obtained with the following formula:

$$\text{Time constant } \tau = R \cdot C$$

3. *Experiment.* The experiment is very simple: Construct the RC circuit, excite, measure the output, and see if the obtained results match the analytical results.

If $R = 1$ kΩ and $C = 1$ μF, then

$$f_c = \frac{1}{2 \cdot \pi \cdot R \cdot C} = \frac{1}{2 \cdot \pi \cdot 1000 \cdot 0.000001} = 159\,\text{Hz}$$

$$\text{Time constant } \tau = R \cdot C = 1000 \cdot 0.000001 = 1 \text{ ms}$$

We excite this RC circuit with a 10-V_{pp} and 159-Hz sinusoid and we measure the rms voltage, as is shown in Figure 3.63.

Reading the input and output rms voltages, we can calculate

$$\text{Output}\,\% = \frac{2.393}{3.413} = 70\%$$

Thus, at least for this RC circuit, the expression of the cut-off frequency is correct.

To experiment with the time constant, we have to excite the previous RC circuit with a 10-V_{pp} and 100-Hz square wave (Figure 3.64).

The analysis of the previous graphic is somewhat complicated. The excursion of the output voltage is from –5 V to +5 V, i.e., 10 V, and 63% of this excursion is 6.3 V. Thus, we must find when the output is 1.3 V (i.e. 1.3 = –5 + 6.3) and read at what time this value is obtained.

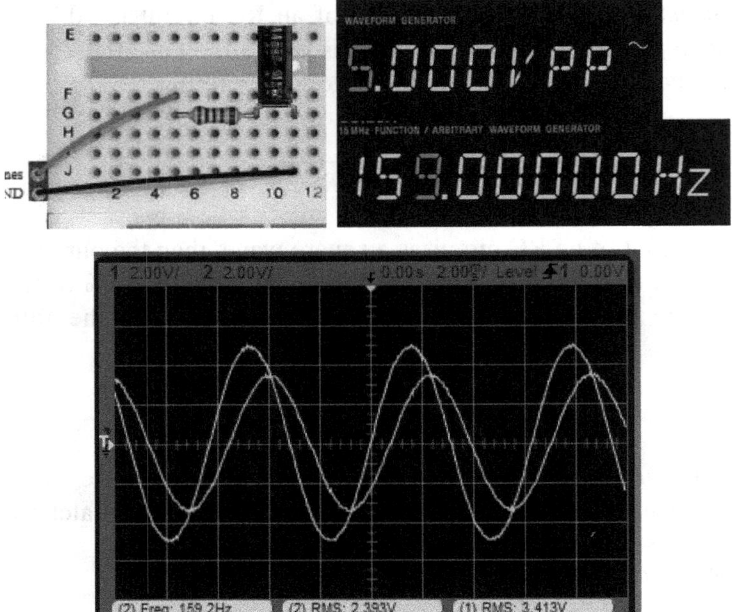

Figure 3.63. Cut-off frequency for an RC low-pass filter.

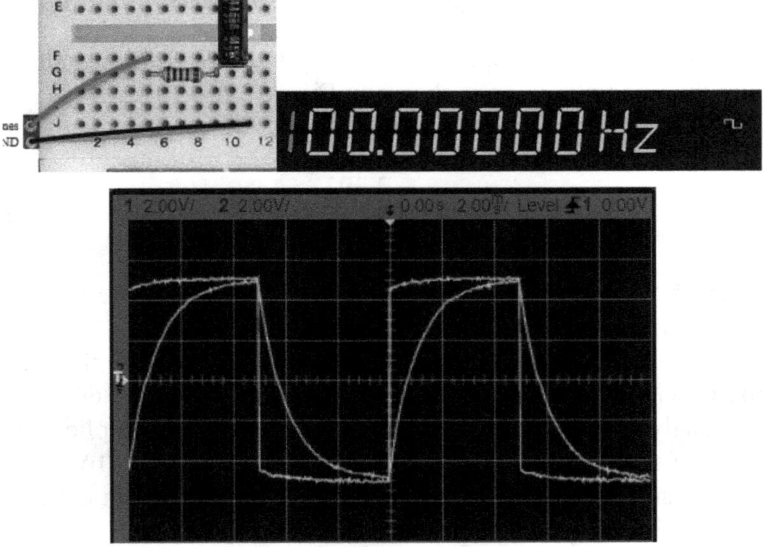

Figure 3.64. Time constant for an RC circuit.

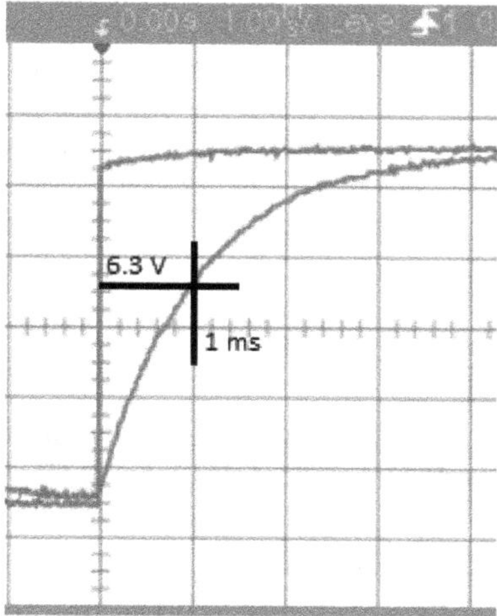

Figure 3.65. Time constant measurement in the output waveform.

In an oscilloscope, there are cursors that can be moved along the visualized waveforms directly to read the values. However, in the VISIR HTML5 version, the cursors are still not implemented, so we must do it ourselves (Figure 3.65).

The obtained value is 1 ms, exactly the expected value. Therefore, the mathematical expression is correct.

4. *Analysis and conclusion.* The mathematical expressions are correct. It is even more important that if we do not know the circuit but the output appears to be an RC circuit, then we can model the circuit with f_c and/or τ. That is, if we measure these two values, we can obtain R and C, and then with the modeled circuit we can design on paper/computer.

Additionally, it is interesting to perform more experiments with other R and C values.

3.3.8 *AC Practice: Using the Cut-off Frequency and Time Constant to Design Circuits*

1. *Introduction.* The cut-off frequency and the time constant are used to design circuits while estimating the behavior of the output. For instance, we can predict at what frequency the noise will be 90% filtered.

2. *Mathematical model.* If we have an RC circuit, analytically, the following expressions describe the filtered % of the input in terms of f_c:

$$50\% = 2 \cdot f_c, \quad 80\% = 5 \cdot f_c, \quad 90\% = 10 \cdot f_c$$

If we have an RC circuit, analytically the following expression indicates how much time is needed to charge a capacitor in terms of τ:

A capacitor is fully charged after $5 \cdot \tau$ s

3. *Practice.* We excite an RC circuit with sinusoidal input waveforms of different frequencies and we measure the rms output voltage. If $R = 1$ kΩ and $C = 1$ μF, then $f_c = 159$ Hz, so we test the circuit with 320 Hz, 800 Hz, and 1600 Hz. Table 3.22 shows that the expression is correct.

Figure 3.66 shows the results for 320 Hz and 1600 Hz. Note that the scales for the input and output are different.

As the time constant is 1 ms, the experiment consists in exciting the RC circuit with a square wave to observe whether after 5 ms the output has reached the V_{max}. The frequency input is 50 Hz.

Table 3.22. Low-pass filter design.

	320 Hz	800 Hz	1600 Hz
$V_{i,rms}$	3.536	3.408	3.493
$V_{o,rms}$	1.451	0.654	0.351
% filtered	59	81	90

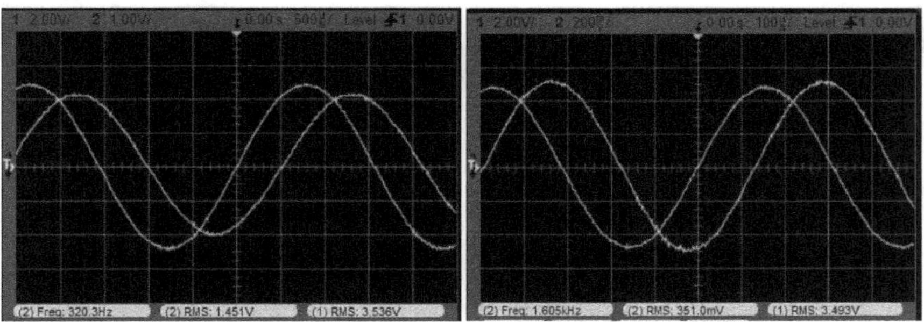

Figure 3.66. Filtering for 320 Hz and 1600 Hz.

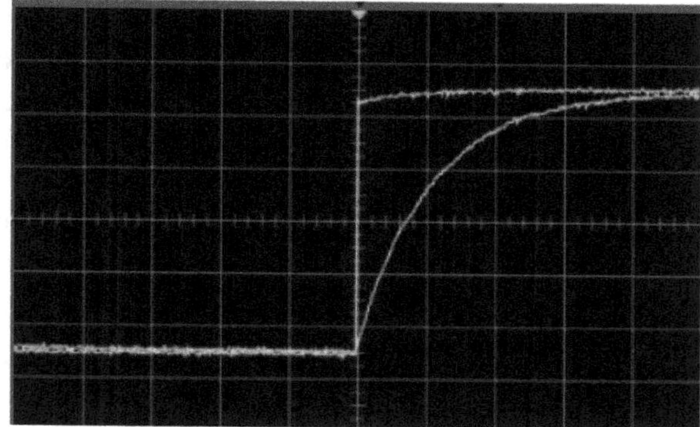

Figure 3.67. Charging curve for a 1-μF capacitor for 50-Hz input.

Looking at Figure 3.67, we see that after 5 ms, the capacitor is totally charged with 5 V.

4. *Analysis and conclusion.* Bearing in mind previous results, if we expect the noise frequency to be 1–10 kHz, then with an RC circuit with 1 kΩ and 1 μF, noise will be reduced by approximately 90%. If R = 10 kΩ and C = 10 μF, then the reduction will even be higher, but at the same time, the non-desired effect upon the input (not upon the noise) will be also greater.

If we have a 1-μF capacitor and we want to charge it in less than 0.5 ms, then the resistor will be smaller than 500 Ω. At the same time, however, the capacitor will be discharged after the same period of time.

Additional practices can be performed with other R, C, and f values.

3.3.9 *AC Experiment: Characterization of RL Circuits*

After the experiments with RC circuits, we can imagine performing the same experiments with RL circuits. We simply replace the capacitor with an inductor.

However, this is not easily done because inductors create problems for the VISIR remote lab. The main problem is that the VISIR has a multiplexor to allow different students to perform experiments at the same time. This is one of VISIR's best characteristics: many (max. 60) students can access it at the same time. This multiplexing time is typically around 300 ms (in AC experiments; see Swartling *et al.*, 2012). Therefore, every 300 ms, the relays are opened and closed, so there are current peaks with high amperage that can damage the VISIR matrix. This time, however, 300 ms, depends on the specific circuit or experiment with the instrumentation.

From our (real) experience, we advise VISIR administrators not to implement RL circuits, at least if they are not experts on this situation.

3.4 Circuits with Diodes

The set of experiments and circuits is as follows:

- Experiment with diode circuit: Characteristic curve
- Experiment with diode circuit: Transfer curve
- Experiment with diode circuit: Threshold diode voltage
- Practice with diode circuit: Half-wave rectifier
- Practice with diode circuit: AC/DC converter, half-wave rectifier + capacitor
- Practice with diode circuit: Characterization of an AC/DC converter
- Experiment with diode circuit: Zener diode characteristic curve
- Practice with diode circuit: Voltage regulator
- Experiment with diode circuit: Different diodes.

3.4.1 *Experiment with Diode Circuit: Characteristic Curve*

1. *Introduction.* A diode is a nonlinear electronic device that is very useful in many simple circuits.

If we want to know how a device behaves, the first step should be to obtain the characteristic curve or *V–I* curve.

A diode has two terminals, anode (A) and cathode (K), and its symbol is shown in Figure 3.68.

2. *Mathematical and logical models.* A diode is a nonlinear device, and its logical behavior is as follows:

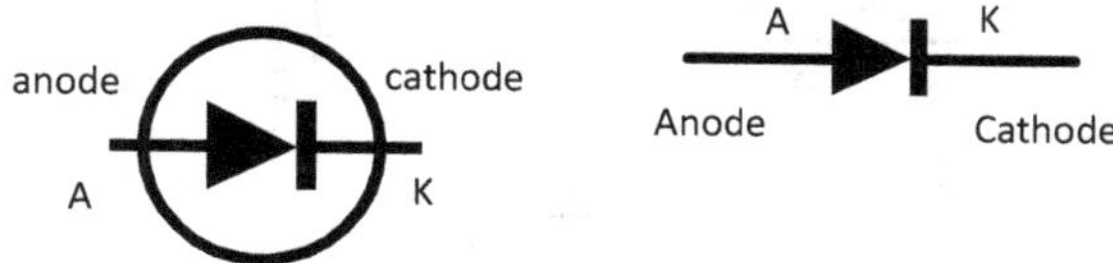

Figure 3.68. Diodes.

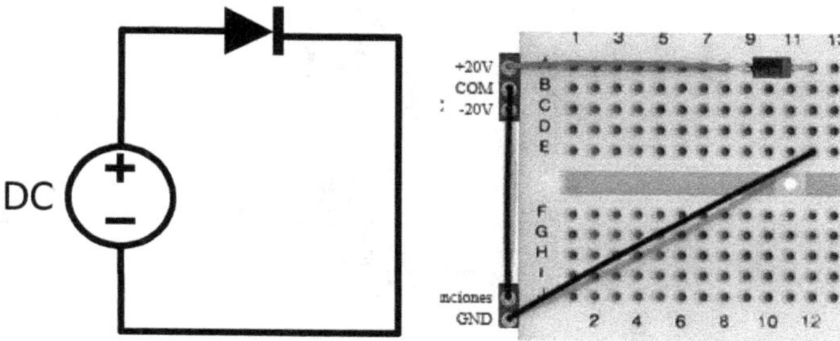

Figure 3.69. Basic diode circuit.

$$\text{if } V_{AK} > V_\gamma \text{ then } I_D > 0 \text{ ON}$$

$$\text{but if } V_{AK} < V_\gamma \text{ then } I_D = 0 \text{ OFF}$$

The values of V_γ and I_D will be obtained in the experiment.
The previous logical model can be expressed with Shockley's model

$$I = I_S \cdot \left(e^{\frac{q \cdot V_D}{n \cdot k \cdot T}} - 1 \right)$$

There is also an electrical model of the diode that will be presented during the analysis.

3. *Experiment.* We simply construct the circuit in Figure 3.69.

As the VISIR has a current limitation of 100 mA, we suggest adding a resistor to limit the current (remember Ohm's law) and to increase the range of V_{AK}. The resistor can be 6 Ω and 1 kΩ, and with the latter, the voltage input can be higher than without the resistors.

Table 3.23. Measurements for the diode characteristic curve.

V	−1	−0.8	−0.6	−0.4	−0.2	0	0.2	0.4	0.6	0.8	1
I	−0.01	−0.01	−0.01	−0.01	−0.01	−0.01	−0.01	0.02	2.6	33.7	85.7

Figure 3.70. Experiment for obtaining the diode characteristic curve.

To obtain Table 3.23, simply open the DC power and excite the circuit with a voltage from −1 V to +1 V. To generate negative voltages we need to change the input from +25 V to −25 V (Figure 3.70).

In Deusto's VISIR, we do not allow more than 100 mA, so the input voltage range is small. If we add a 6-Ω resistor (Figure 3.71), the voltage range input can be higher (Table 3.24).

Figure 3.72 evidences in graphic fashion the diode's behavior.

4. *Analysis and conclusion.* The behavior of the diode is nonlinear. In simple terms, it is ON when V is positive and it is OFF when V is negative.

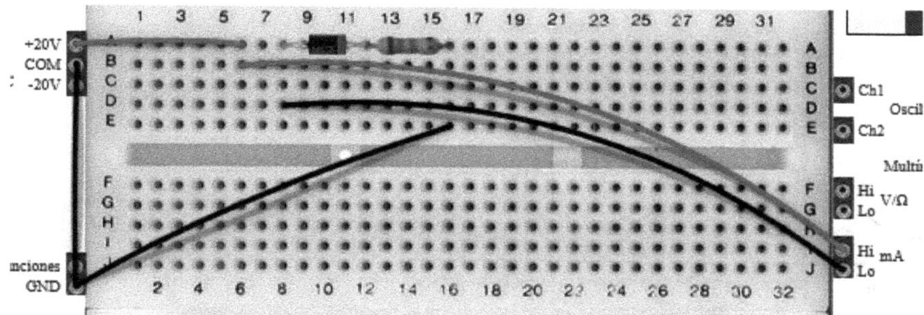

Figure 3.71. Experiment to obtain the diode characteristic curve with a 6-Ω resistor.

Table 3.24. Measurements for the diode characteristic curve with a 6-Ω resistor.

V	−1.6	−1	0	0.2	0.4	0.6	0.8	1	1.2	1.4	1.6
I	−0.01	−0.01	−0.01	−0.01	0.01	1.8	13.6	31.6	51.2	71.6	91.5

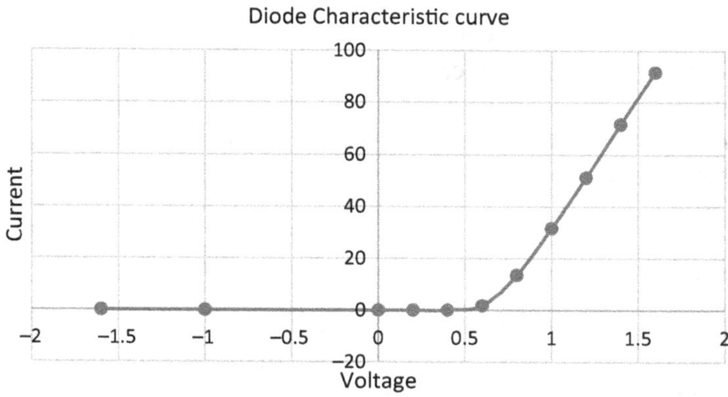

Figure 3.72. Experimental diode characteristic curve.

When it is ON, the current is positive, and when it is OFF, the current is 0. The diode is similar to a switch.

In more detail, the diode is ON when V is higher than 0.6–0.7 V. This voltage value is the threshold, V_γ.

Observing the ON behavior, we see that the current grows dramatically with a high slope, so there is a low resistance in the diode. Observing Table 3.24, we note that r is around 10 Ω, and as there is 6 Ω, then the

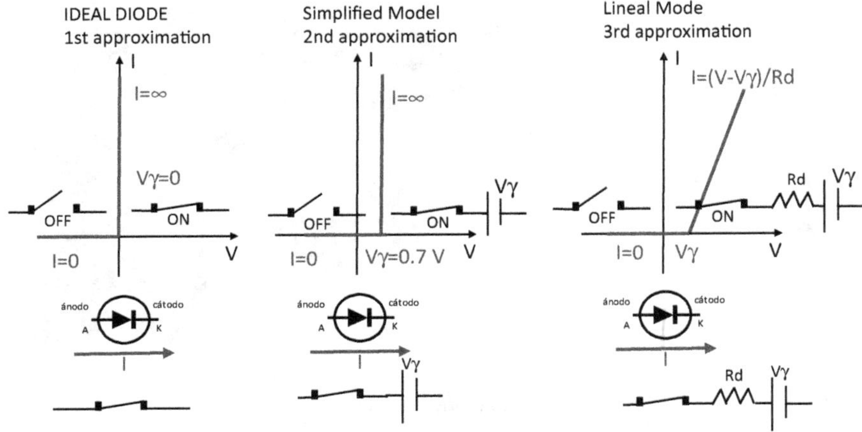

Figure 3.73. Electrical models for diode.

diode resistance, r_d, is about 4 Ω. Thus, we can simplify even further the electrical model of the diode with $r_d = 0$ Ω. Figure 3.73 shows the three approximations to the electrical model of the diode.

Using the 1-kΩ resistor instead of the 6-Ω, we can increase the input range voltage to 5–10 V. With this, we can see that in the ON side, there is a linear relation between *V* and *I*.

3.4.2 *Experiment with Diode Circuit: Transfer Curve*

1. *Introduction.* In the previous experiment, we determined the characteristic curve of the diode or *V–I* curve. In this new experiment, we will relate input to output voltage or V_i–V_o curve.

The characteristic curve refers to the diode, while the transfer curve refers to the circuit. This means that different circuits have different transfer curves.

2. *Mathematical and logical model.* The experiment will test the circuit in Figure 3.74.

As we have seen, the diode has two different behaviors, ON and OFF, so there are two mathematical solutions to the circuit.

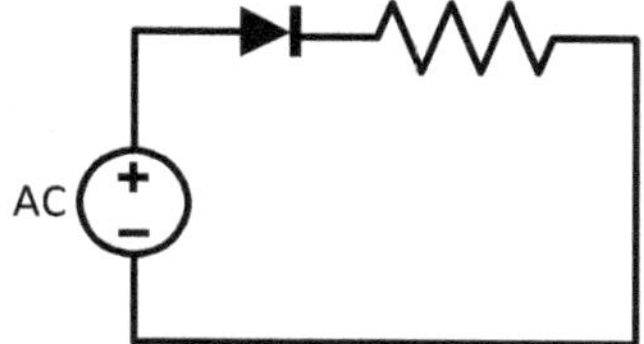

Figure 3.74. Diode transfer curve circuit.

Table 3.25. Voltage measurements for the diode transfer curve.

V_i	−2	−1	0	1	2	3	4	5	6	7	8	9	10
V_o	0	0	0	0.4	1.4	2.4	3.4	4.3	5.3	6.3	7.3	8.3	9.3

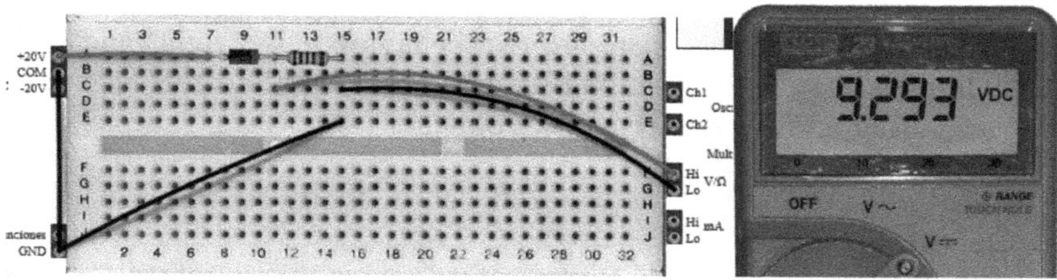

Figure 3.75. Diode transfer curve circuit in VISIR.

ON: If $V_i > V_\gamma$, then applying Ohm's and Kirchhoff's laws

$$V_o = V_i - V_\gamma - I_D \cdot r_d, \text{ but if } V_\gamma = 0\,\text{V} \text{ and } r_d = 0\,\Omega, \text{ then } V_o = V_i$$

OFF: If $V_i < V_\gamma$, then the circuit is open and $I_D = 0$, so $V_o = 0\text{V}$

3. *Experiment.* The circuit under experiment is a series of the diode and a 1-kΩ resistor. We use this resistor instead of the 6-Ω resistor because we simply want the transfer curve. We excite the circuit with −2 V to +10 V and we measure V_o (Table 3.25 and Figure 3.75).

The transfer curve is in Figure 3.76.

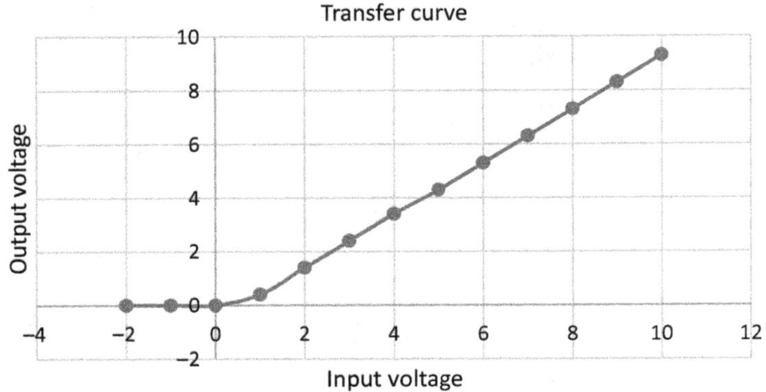

Figure 3.76. Experimental diode transfer curve circuit.

We can again see that there are two behaviors: ON (right) and OFF (left). In the ON side, the relation between V_i and V_o is linear. If we adjust this part, using a linear regression, we obtain

$$V_o = V_i - 0.7\ \text{V}$$

4. *Analysis and conclusion.* The effect of a diode in this simple circuit has two behaviors:

$$\text{ON: If } V_i > 0\ \text{V}, \quad \text{then } V_o = V_i - 0.7\ \text{V}$$
$$\text{OFF: If } V_i < 0, \text{ then } V_o = 0\ \text{V}$$

0.7 V is the threshold voltage of the diode, V_γ.

Additionally, we can repeat the experiment with a 10-kΩ output resistance.

3.4.3 *Experiment with Diode Circuit: Threshold Diode Voltage*

1. *Introduction.* In the two previous experiments, we have seen there is voltage drop in the diode. This voltage drop is explained by the semiconductor theory, but at least we can measure and characterize it.

2. *Mathematical model.* In the first experiment 3.4.1, we marked that the threshold was around 0.6–0.7 V. We obtained this value by looking at the voltage in which the current starts to be higher than 0.

In the second experiment, we have seen that the difference between V_i and V_o was around 0.7 V.

However, these two previous statements are only true when the diode is ON. What happens when the diode is OFF? In this situation, as the current is 0, then $V_o = 0$, so $V_d = -V_i$ because $V_o = V_i + V_d$ (Kirchhoff's second law).

3. *Experiment.* The experiment is exactly the same as the previous one, but, in this case, we measure the voltage drop (Figure 3.77) in the diode, not in the output (Table 3.26 and Figure 3.78).

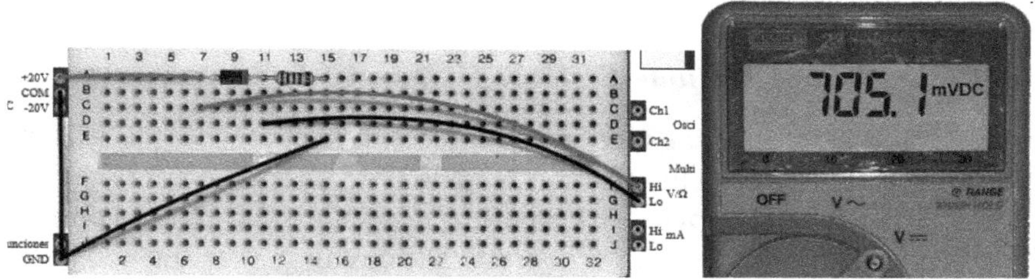

Figure 3.77. Diode threshold voltage.

Table 3.26. Threshold voltage measurements.

V_i	−2	−1	0	1	2	3	4	5	6	7	8	9	10
V_d	−2	−1	0	0.56	0.62	0.63	0.65	0.66	0.67	0.69	0.69	0.7	0.71

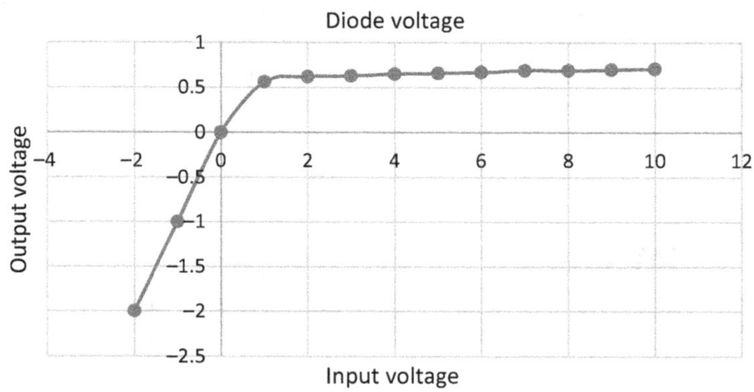

Figure 3.78. Diode threshold voltage curve.

4. *Analysis and conclusion.* When the diode is ON, V_d is constant and around 0.7 V (at least for this diode, 4N1002). This value is the threshold voltage.

When the diode is OFF, V_d is equal to $-V_i$. That is, the diode supports all the input voltage between its terminals. If this input voltage is too high, the diode can be damaged.

If we look at the graph carefully, we can see that the voltage increases slowly when the diode is ON. This is because if V_i grows, the current grows, and the voltage drop in the internal resistance (r_d) grows:

$$\text{ON}: V_d = V_\gamma + I_d \cdot r_d$$

3.4.4 *Practice with Diode Circuit: Half-wave Rectifier*

1. *Introduction.* One of the most popular electronic circuits is the AC/DC converter (Figure 3.79). Its function is to convert alternating current into direct current. For instance, the current of the plug is AC but the current in the computer is DC, so we need an AC/DC circuit to plug the computer into a domestic AC socket.

2. *Mathematical and logical models.* Any AC signal has an average value equal to 0 V because half of the wave is above 0 V and the other half is below 0, so the sum of the two halves is always 0. The signal form (sinusoidal, triangular, etc.) is not important. A DC signal has a constant value, which is the average. If we have a 5-V DC power source, the constant value is equal to the average: 5 V.

At the same time, if we remember the diode's behavior, we know that if the input is positive, the output equals the input (minus the threshold voltage), and if the input is negative, the output is equal to 0 V.

Thus, the output has always a positive value, and so the average will not be 0 V.

Figure 3.79. AC/DC commercial converter of a laptop.

$$V_{\mathrm{DC}} = V_{\mathrm{average}} = \frac{\int_0^T v(t)dt}{T}$$

In the case of a half-wave rectifier, the mathematical model is

$$V_{\mathrm{DC}} = V_{\mathrm{average}} = \frac{V_{\mathrm{max}}}{\pi}$$

3. *Practice.* The circuit is very simple: An AC power source connected with a diode and finally connected to an output resistor (Figure 3.80).

The output resistance is 1 kΩ, and the input is a 10-V_{pp} and 1-kHz sinusoid.

In Figure 3.81, we observe the output signal and measure its frequency and average and maximum voltages.

Figure 3.82 shows the output only.

There is a difference between output and input, around 0.8 V, the threshold voltage of the diode.

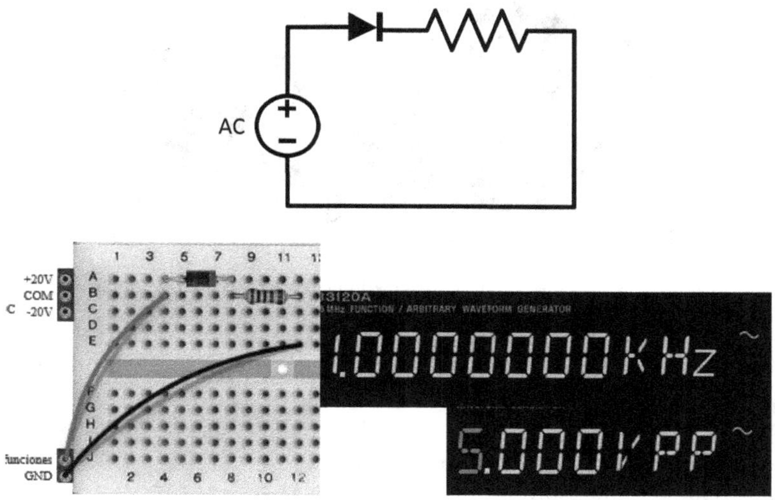

Figure 3.80. Half-wave rectifier circuit and VISIR measurements.

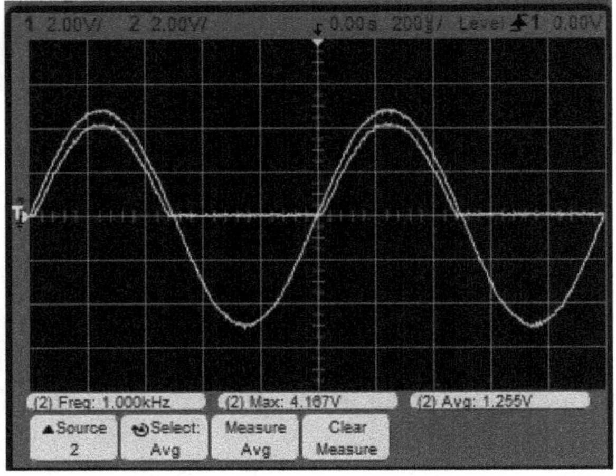

Figure 3.81. Half-wave rectifier voltage in VISIR oscilloscope I.

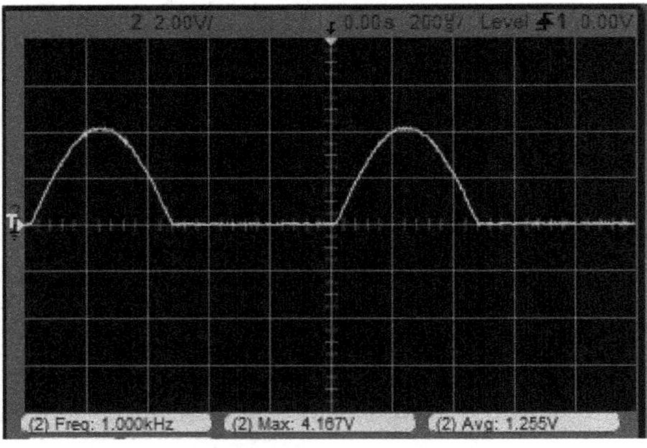

Figure 3.82. Half-wave rectifier voltage in VISIR oscilloscope II.

4. *Analysis and conclusion.* If we analyze the previous images and values:

- The output is half of the input wave.
- The average output voltage is not 0 V, it is 1.255 V.
- The output is not flat, it is not like a DC signal.

Looking at the mathematical model, we see that, approximately, the expression is correct.

$$V_{DC} = V_{average} = \frac{V_{max}}{\pi} = \frac{4.17}{\pi} = 1.46\,\text{V and the measure is } 1.26\,\text{V}$$

The half-wave rectifier is only the first step in the AC/DC conversion. The second part will be the next practice.

The analysis of the expression should be repeated for different frequencies and voltage inputs.

What happens if we reverse the diode: The cathode (K) connected to the function generator and the anode connected to the output resistor?

3.4.5 *Practice with Diode Circuit: AC/DC Converter, Half-wave Rectifier + Capacitor in Parallel*

1. *Introduction.* In the previous practice, we have seen that when using a diode we can obtain a half-wave sinusoid with a non-0 V average voltage. However, the output signal is not a DC signal; it is not "flat".

Using a capacitor in parallel with the output resistor, we obtain a basic AC/DC converter.

2. *Mathematical and logical model.* Previously, we studied the time constant of a capacitor. We saw that the capacitor is charged if there is a current flowing through it, and when there is not any current in the circuit, the current flows from the capacitor.

Logically, when the input is positive, the capacitor is charged because the diode is ON, but when the diode is OFF, even though the output resistor is isolated from the input, the output receives current from the capacitor, until it is discharged.

The following mathematical expression provides the average voltage of this AC/DC basic circuit.

3. *Practice.* In this case, we only need to add a capacitor, 1 μF, in parallel with the output resistor, 1 kΩ (Figure 3.83).

The oscilloscope shows in Figure 3.84 that now the output is more or less flat. It has a clear average value and it has an oscillation (sawtooth) around the average.

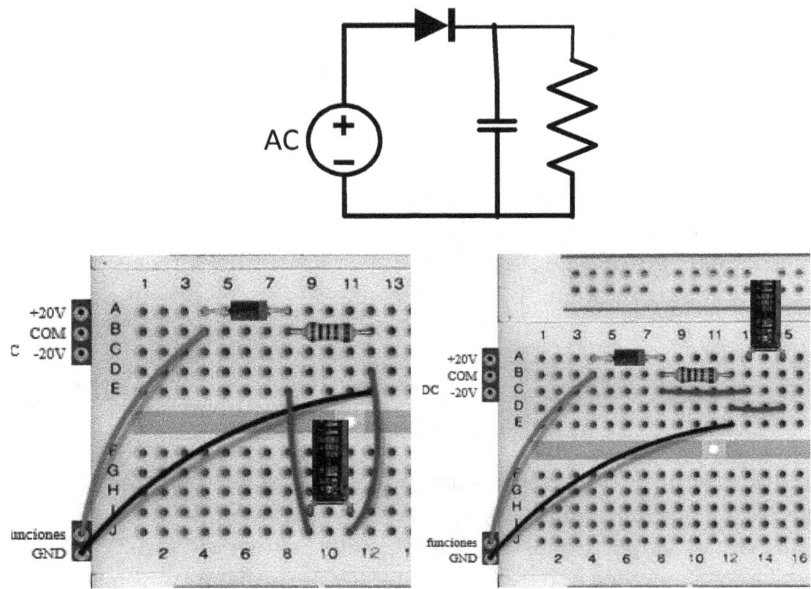

Figure 3.83. Half-wave rectifier + C filter.

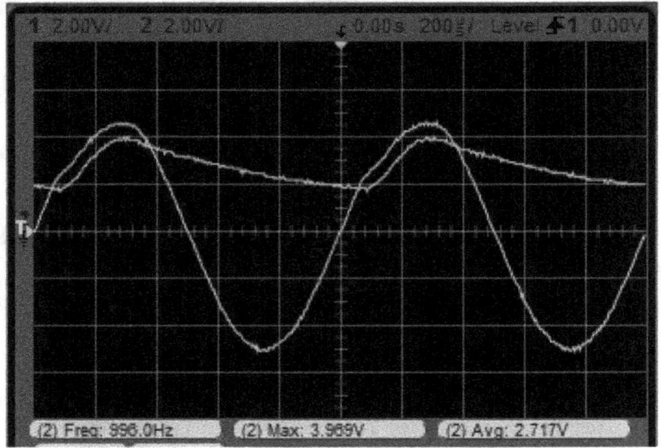

Figure 3.84. Half-wave rectifier + C filter waveform I.

We can also see that the diode is ON for a shorter time than before. It is ON when the capacitor charges, and this is because now V_{AK} is $V_{in} - V_{cap}$.

This effect can be seen more clearly if the input is removed from the oscilloscope, as is shown in Figure 3.85.

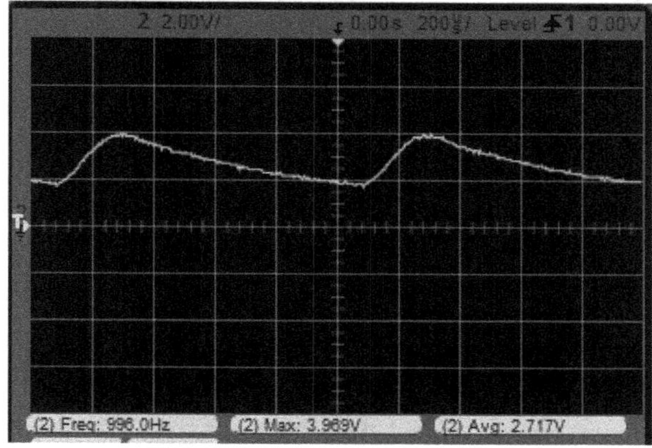

Figure 3.85. Half-wave rectifier + C filter waveform II.

Looking at the graphic, we can see that the average is, more or less, 3 V. The oscilloscope indicates that the value is 2.72 V.

In Figure 3.86, we see that when the diode is ON, the capacitor charges, and when the diode is OFF, the capacitor discharges and maintains the output voltage value at a value that decreases with time (time constant).

If we replace the 1-μF capacitor with a 10-μF one, we obtain the result shown in Figure 3.86.

Is the output of this circuit a DC signal? What is the DC value? Around 3 V?

At this moment, we can use the AC mode of the oscilloscope. In this mode, we see the voltage minus the average, i.e., we see only the alternant part. To obtain a clear image, we need to change the voltage scale. It was 2 V, and now it is 100 mV (Figure 3.87).

Now we see in Figure 3.88 that there is a 150-mV AC "triangular" wave superimposed on the 3-V DC. At this moment, we can measure the rms value to characterize the alternant part of the output voltage of the AC/DC converter.

4. *Analysis and conclusion.* We have seen that this simple circuit is an AC/DC converter; now instead of a sinusoidal signal, we have, more or less, a DC signal.

Now we should analyze the "more or less". We need to characterize an AC/DC converter. The second circuit, with a 10 μF, is better than the first one, with 1 μF. How much better is it?

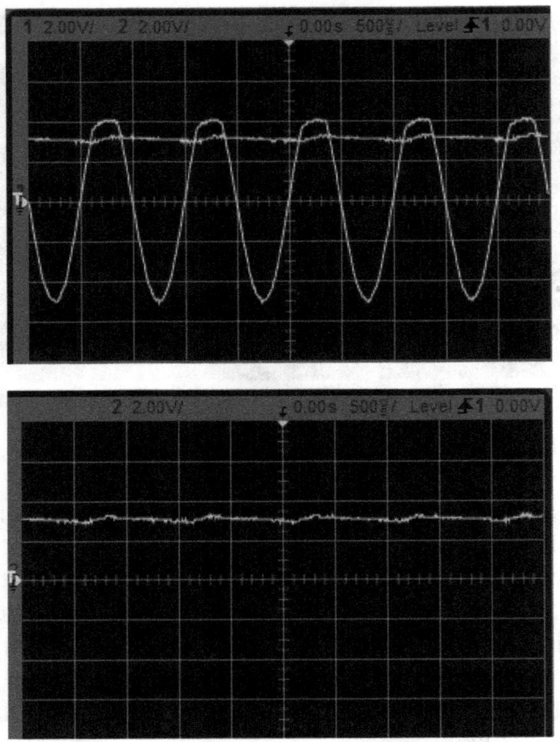

Figure 3.86. Half-wave rectifier + 10 μF C filter waveform.

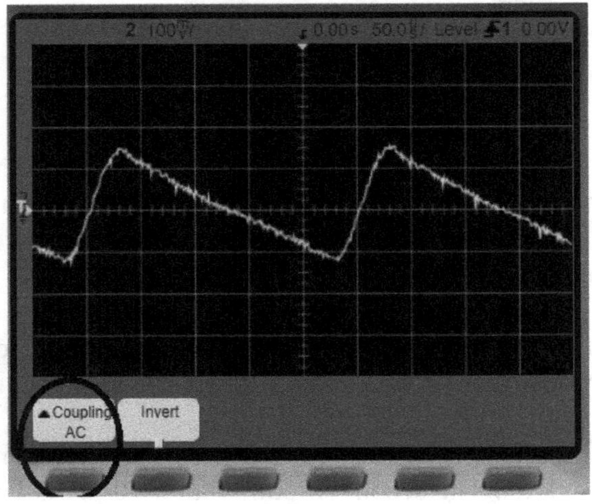

Figure 3.87. AC coupling mode in the VISIR oscilloscope.

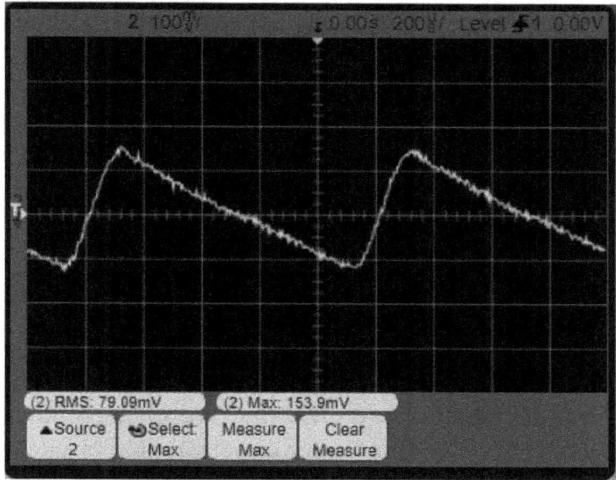

Figure 3.88. Detailed AC half-wave rectifier + 10 μF C filter waveform.

3.4.6 *Practice with Diode Circuit: Characterization of an AC/DC Converter*

1. *Introduction.* In the previous practice, we have seen that a very simple circuit has converted an AC signal into a DC signal.

We have seen that the quality of the AC/DC conversion depends on the capacitance of the capacitor: The higher the capacitance, the higher the quality of the DC output.

Therefore, the capacitor affects the AC/DC converter, but there are more variables in the circuit: Frequency of the sinusoid, output resistance, etc. This practice consists of varying these variables and observing their effects in the AC/DC converter.

2. *Mathematical model.* It is obvious that the DC output obtained with $C = 10\ \mu\text{F}$ is better than the obtained output with $C = 1\ \mu\text{F}$, but why? Can we express this with numbers?

There are some indicators that are used by designers: form factor (FF) and ripple factor (RF).

$$\text{FF} = \frac{V_{\text{rms}}}{V_{\text{cc}}} \quad \text{RF} = \frac{V'_{\text{rms}}}{V_{\text{cc}}}$$

V'_{rms} is the rms voltage value of the alternating part of the output. That is, after the conversion, the output has a constant part (DC part) and a variable part (AC part), so $v_o(t) = V_{DC} + v_{AC}(t)$. The use of ′ represents the alternating part.

Ideally, FF should be 100% because this means that all the output (measured with V_{rms}) is DC. In the worst case, V_{cc} is 0, and the FF is ∞%.

Ideally, RF should be 0% because this means that all the output is DC, and the alternating part is 0. In the worst case, RF is 100% because this means that all the output is AC.

3. *Practice.* The process is very simple and it is a typical experiment: we fix all but one independent variables, and we change the value of one single variable, i.e., the variable under experiment. For each experiment, we measure V_{cc}, V_{rms}, and V'_{rms} to obtain FF and RF. The input is a 5-V_{max} sinusoidal wave of 100 Hz (Figure 3.89).

Before starting, we explain how we can measure V'_{rms}. Figure 3.90 shows that if we click on the oscilloscope channel number 2, then a new control appears in the graphic.

If we click on the bottom left button, we can change DC to AC. After this, if we again click on Perform Experiment, we obtain Figure 3.91. In this new graph, the output signal is centered, and VISIR has subtracted the DC value from the signal, so we only see the alternating signal, the AC part of the output. If we measure the signal, the obtained values will be AC values. If we want to obtain AC and DC values, we must change from AC to DC mode in the VISIR oscilloscope (Figure 3.90).

In the first scenario, we maintain the circuit ($R = 1$ kΩ and $C = 1$ μF), and we change the input frequency. Table 3.27 shows the results for the different frequencies.

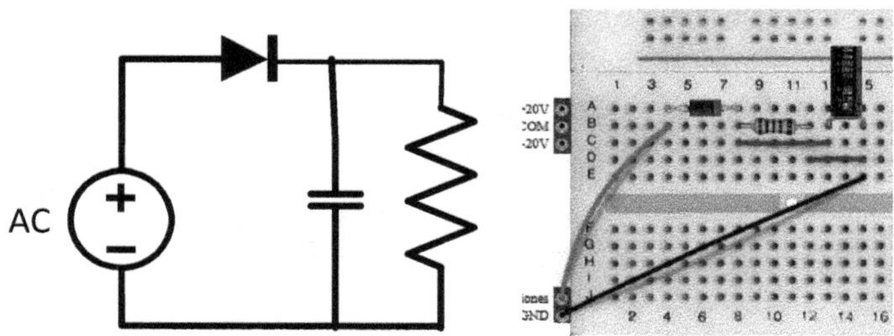

Figure 3.89. Half-wave rectifier + C filter.

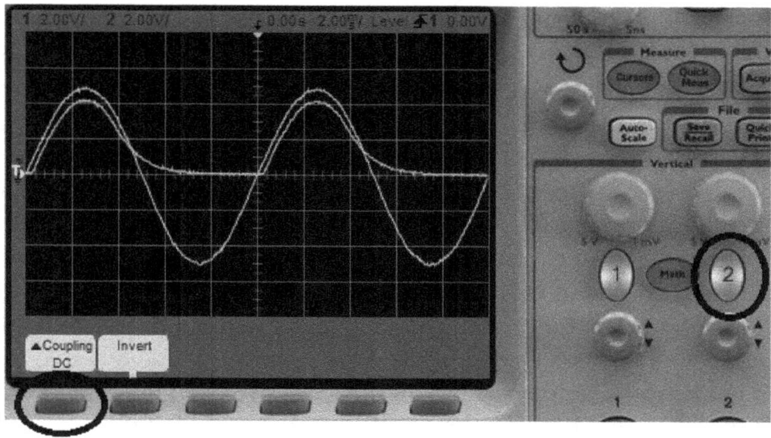

Figure 3.90. Half-wave rectifier + C filter waveform in VISIR oscilloscope.

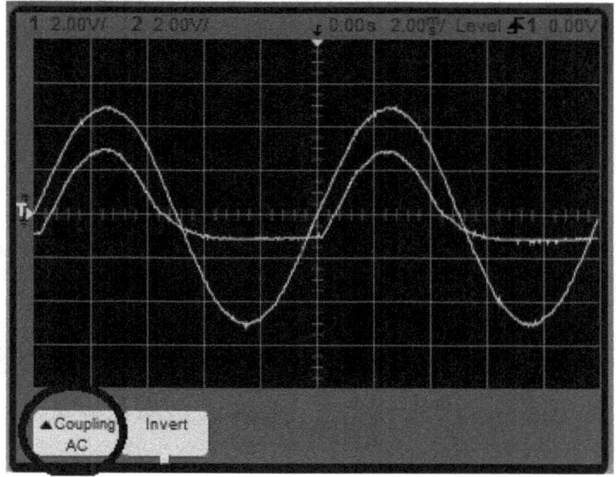

Figure 3.91. AC Half-wave rectifier + C filter waveform in VISIR oscilloscope.

Table 3.27. Form factor and ripple factor of an AC/DC converter.

	100 Hz	200 Hz	500 Hz	1000 Hz	2000 Hz
V_{rms}	2.1	2.2	2.5	2.8	3
V_{cc}	1.4	1.6	2.3	2.7	3
V'_{rms}	1.46	1.33	0.97	0.61	0.34
FF (%)	150	137	110	104	100
RF (%)	114	83	42	23	11

It is clear that the DC output is better when the frequency is high. The RF at 100 Hz was 114% and at 1000 Hz was 23%, and the FF at 100 Hz was 150% and at 1000 Hz was 100%. Figure 3.92 shows the output at 100 Hz and 1000 Hz.

Logically, if the time between the capacitor's charges is shorter, the discharging time will also be shorter, and then the output will be higher.

In the next scenario, we maintain the frequency (500 Hz) and we change *R* and *C*. We measure the AC/DC converter quality with FF and RF, first and second in each cell of Table 3.28.

For $R = 1$ kΩ and $R = 10$ kΩ at 500 Hz, we can see the two outputs in Figure 3.93.

What happens to the current? In this case, we need the multimeter to measure the current because the oscilloscope only measures voltage. It is important to note that when the multimeter is in AC, it measures the rms value of the signal (Figure 3.94), voltage, or current, but when the oscilloscope is in AC, it measures only the alternating part. At 100 Hz, the output current is 1.5 mA, but at 1000 Hz, the output current is 0.66 mA.

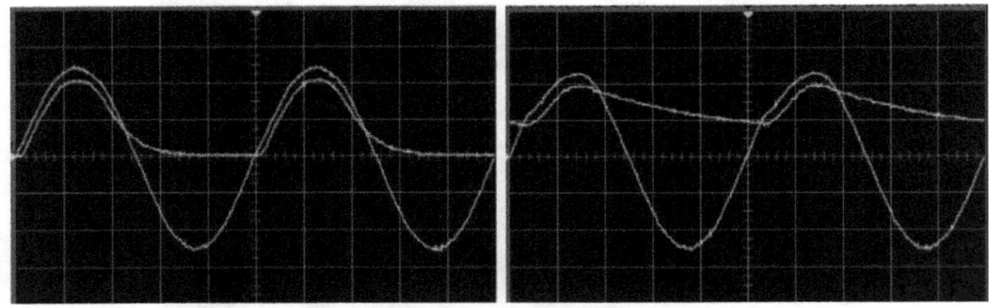

Figure 3.92. AC/DC converter waveforms for 100 Hz and 1000 Hz.

Table 3.28. Form factor and ripple factor of an AC/DC converter with different capacitors.

	1 μF (%)	10 μF (%)
1 kΩ	109	100
	42	6.4
10 kΩ	100	
	4.3	

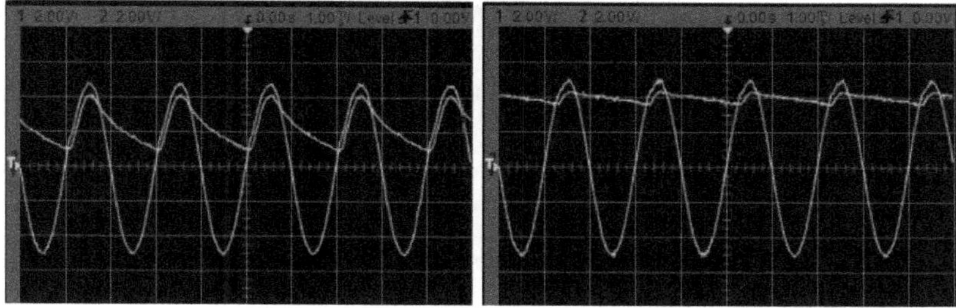

Figure 3.93. AC/DC converter waveforms for 1-kΩ and 10-kΩ resistors.

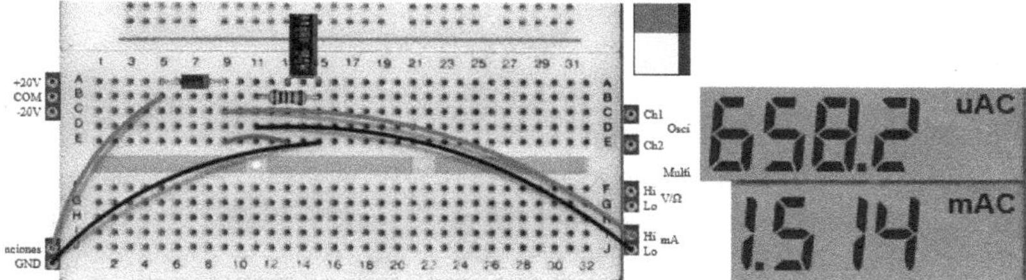

Figure 3.94. Current measurement in an AC/DC converter.

So at what frequency do we have the best performance? At 1000 Hz, because the RF is lower, or at 100 Hz, because the output current is higher?

4. *Analysis and conclusion.* There are two main conclusions that can be obtained from the previous results:

- The frequency, resistance, and capacitance affect the performance of the AC/DC converter.
- The RF value is better than the FF value to measure the converter's performance.

In general, the input frequency and the output resistor are fixed, so the designer can change the capacitor. A high capacitance value appears to be better, but at the same time we can see that V_{cc} decreases, and this may not be good. Furthermore, what happens to the current? If V_{cc} decreases, the output power decreases. At the same time, we can see that if the capacitance is high, the input signal is affected, and this is not good.

As usual, then, there is a balance between different performance criteria.

Remember that to measure different signals, we can use the multimeter and/or the oscilloscope, and sometimes we can even measure a signal using an indirect method. In this case, if we measure the AC voltage in the output using the oscilloscope, we can then divide this voltage by R and obtain the rms value of just the alternating part of the current.

3.4.7 *Practice with Diode Circuit: Efficiency of an AC/DC Converter*

1. *Introduction.* As we have seen in the previous practice, when we design an AC/DC converter, we have to consider not only the RF index but also the current and power. The RF may be good, but the current is too low.

This means that we should pay attention not only to effectiveness but also to efficiency.

2. *Mathematical model.* The efficiency of an AC/DC converter can be measured by comparing the total output power with the DC output power.

$$\text{Efficiency} = \eta = \frac{P_{\text{DC, output}}}{P_{\text{rms, output}}} = \frac{V_{\text{DC}} \cdot I_{\text{DC}}}{V_{\text{rms}} \cdot I_{\text{rms}}}$$

Performance can also be measured as the relation between the power of the input signal and the power of the output signal.

$$\text{Performance} = \frac{P_{\text{rms, output}}}{P_{\text{rms, input}}} = \frac{V_{\text{rms, output}} \cdot I_{\text{rms,output}}}{V_{\text{rms, input}} \cdot I_{\text{rms, input}}}$$

3. *Practice.* To obtain efficiency and performance, we need to measure different values of voltages and currents. How can we do this in the VISIR?

The electronic instrumentation in the VISIR, and not only in the VISIR, has some particularities that need to be known:

- In the VISIR, the multimeter in DC voltage mode measures the instant value of the voltage. So if the measured signal is constant (DC voltage), then the measured value will be correct. But if we measure an AC voltage signal, then the measured value will not be correct because the instant value will not be the average value. In Figure 3.95, we can see two different DC values for the same AC wave, which is not correct. The same situation occurs with current.

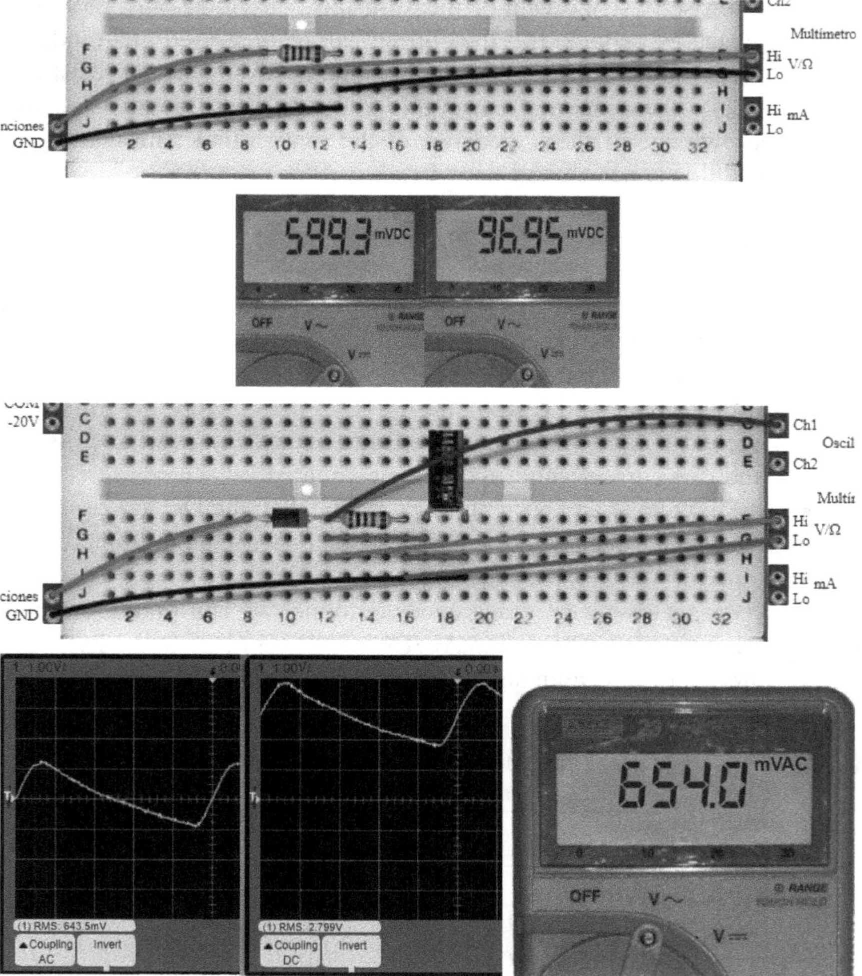

Figure 3.95. Indirect measurements method in VISIR.

- In the VISIR multimeter, and in general in all multimeters, the AC mode measures the rms value of the measured AC signal, removing the DC part. That is, if the measured signal has AC and DC parts, the AC mode only measures the AC part. Figure 3.95 shows that the multimeter measures only the rms AC value, and the oscilloscope measures the rms total value, AC+DC. Some multimeters have the AC+DC label, and this means that they can measure the AC, DC, and AC+DC values.
- The oscilloscope cannot measure the current; it only plots and measures voltage. But the current can be measured in an indirect way (Figure 3.95): If the rms voltage is 2.5 V and the resistor is 1 kΩ, then the rms current is 2.5 mA. This method requires knowledge of the impedance of the branch under measurement, and this is not always known.

To sum up (see also Table 3.29),

- If the voltage and current are pure AC (without DC value), then the rms values can be measured either with the oscilloscope or with the multimeter, and the DC values would be 0.
- If the voltage and current are AC+DC, then the DC and rms voltage value can only be measured with the oscilloscope, and the DC and rms current value should be estimated.
- If the voltage and current are pure DC, then the DC values can be measured either with the oscilloscope or with the multimeter.

In accordance with previous considerations, we can obtain the efficiency and the performance of an AC/DC converter (1 μF and 1 kΩ) with a 5-V_{max} and 1000-Hz input signal (Figure 3.96 and Table 3.30).

Table 3.29. Methods for measuring different current and voltage values.

	V_{rms}	I_{rms}	V_{DC}	I_{DC}
Pure DC	=	=	M/O	M/O
Pure AC	M/O	M/E	0 V	0 A
AC+DC	O	E	O	E

Notes: M: multimeter, O: oscilloscope, E: estimation.

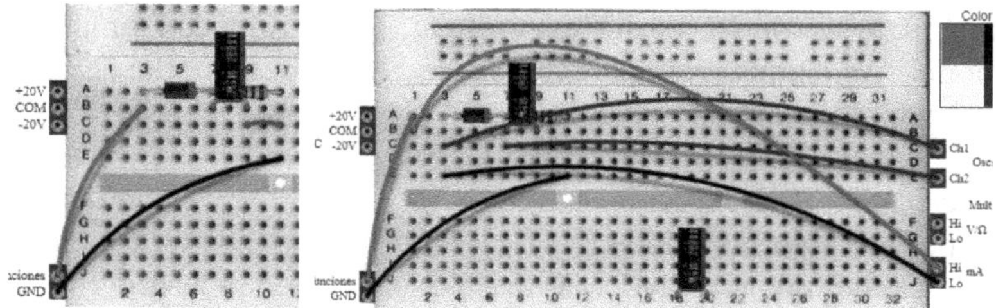

Figure 3.96. Indirect measurements in VISIR for an AC/DC converter.

Table 3.30. Indexes for an AC/DC converter.

$V_{rms,i}$ (O/M)	$I_{rms,i}$ (M)	$V_{rms,o}$ (O)	$I_{rms,o}$ (E)	$V'_{rms,o}$ (O/M)	$I'_{rms,o}$ (E)	$V_{DC,o}$ (O)	$I_{DC,o}$ (E)
3.41 V 3.37 V	5.31 mA	2.84 V	2.84 mA	0.62 V 0.65 V	0.62 mA	2.76 V	2.76 mA

RF	FF	$P_{rms,i}$	$P_{rms,o}$	$P_{DC,o}$	η	Performance
22%	103%	18.11 mW	8.07 mW	7.62 mW	94%	45%

4. *Analysis and conclusion.* We can now compare in VISIR different converters with different inputs, examining effectiveness and efficiency.

To perform this analysis, it is important to know the characteristics of the different electronic instruments in VISIR.

3.4.8 *Experiment with Diode Circuit: Zener Diode Characteristic Curve*

1. *Introduction.* A Zener diode is a semiconductor device similar to a normal diode, but its use and application are different.

To understand what the main characteristic of a Zener diode is, we obtain its characteristic curve: V vs. I.

2. *Mathematical model.* In this experiment, we do not need a mathematical model. We simply excite the Zener diode and measure the current and the voltage in the diode.

3. *Experiment.* This experiment, shown in Figure 3.97, is similar to one developed for a normal diode.

We excite the terminals of the Zener diode, measure the diode current (Table 3.31), and draw the characteristic curve (Figures 3.98–3.100). Before starting, it is important to note that the most important behavior of the diode is on the left-hand side of the graph when the diode voltage is negative. Because of this, the Zener diode is in reverse fashion, with the cathode connected to the power and the anode connected to the ground.

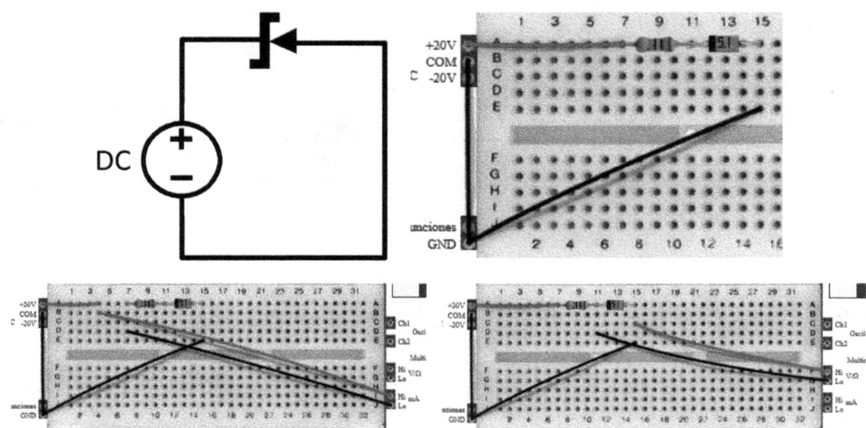

Figure 3.97. Zener diode circuit.

Table 3.31. Measurements for the diode Zener characteristic curve.

V_{DC}	0	–1	–2	–3	–4	–5	–6	–7	–8	–9	–10	–20
I_Z	–0	–0	–0	–0	–0.2	–1	–2	–4	–6	–8	–10	–32
V_Z	0	–1	–2	–3	–4	–4.5	–5.1	–5.2	–5.2	–5.2	–5.3	–5.3

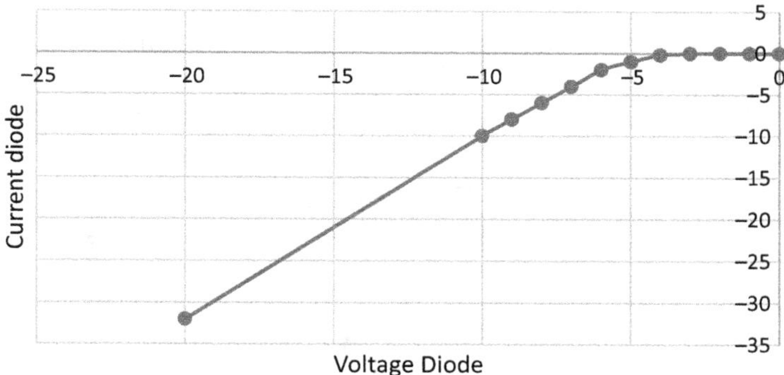

Figure 3.98. Zener diode characteristic curve.

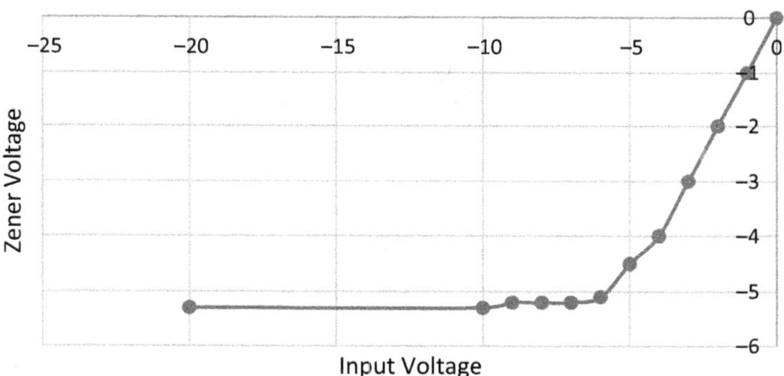

Figure 3.99. Zener diode characteristic curve details.

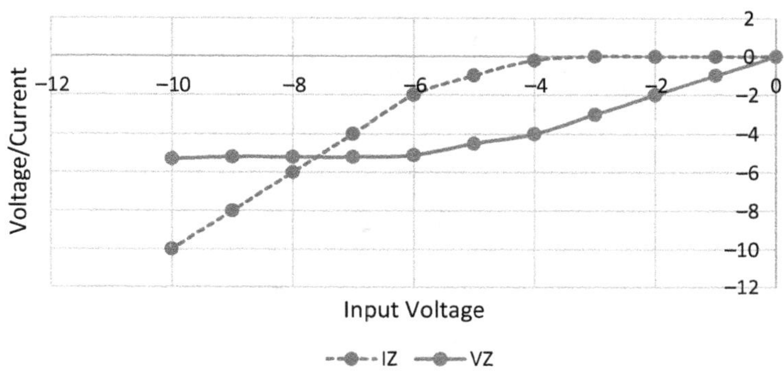

Figure 3.100. Zener diode curve: Current and voltage.

4. *Analysis and conclusion.* Looking at the graphics and results of the experiment, some interesting conclusions can be drawn:

- The Zener diode is used with inverse polarization, in reverse fashion.
- If the input voltage (negative) grows, the current grows.
- If the input voltage (negative) grows, the current grows but the diode voltage remains constant at around 5 V.

It is clear that the Zener diode is very different from the normal diode but is also useful in some electronic circuits. We also recommend completing the experiment with positive input voltages.

3.4.9 *Practice with Diode Circuit: Voltage Regulator*

1. *Introduction.* As we have seen, a Zener diode is able to maintain a voltage even if the current increases. This means that if for any reason the current increases, then this increase can be driven through the diode without affecting the rest of the circuit.

2. *Mathematical and logical model.* In the circuit in Figure 3.101 we have implemented a voltage divider. If the input is 7.5 V, then the output will be around 5 V.

$$I = \frac{V}{R_{\text{tot}}} = \frac{7.5V}{1470\Omega} = 5.10 \text{ mA}, \quad V_o = I \cdot R_o = 5.10 \cdot 1 = 5.10\,\text{V}$$

But what happens if the input voltage is not constant at 7.5 V? Maybe the battery is not very good and its value varies between 7.5 V and 12 V. What happens? Looking at the previous expressions, the output voltage will be between 5 V and 8 V, not the required 5 V.

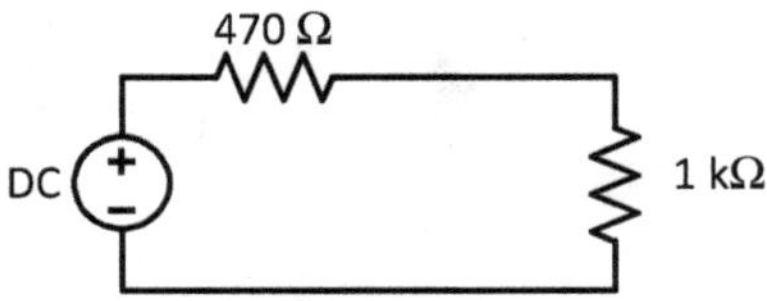

Figure 3.101. Zener diode circuit I.

3. *Practice.* If we add a Zener diode in parallel to the output, we obtain the circuit in Figure 3.102. What we must do is (1) construct the circuit, (2) change the input voltage, and (3) measure the output voltage.

We see that the input value rises from 7.5 V to 12 V, but the output voltage remains around 5 V. Using a Zener diode helps the designer to regulate the voltage.

The same approach can be used in the AC/DC circuit. If we add a Zener diode in parallel with the output resistor (Figure 3.103), the voltage output will be around 5 V, even if the input voltage changes. The AC/DC circuit has $R_o = 1\ k\Omega$, $C = 10\ \mu F$, and an input voltage of 10 V_{pp} at 200 Hz.

In Figure 3.104, we see the output voltage when the input is 7 V (left) and 10 V (right). We can see that the average voltage remains around 4.5 V. Without the Zener diode, the output voltage would have been 4.4 V and 6.5 V, i.e., the input voltage would have affected the output voltage.

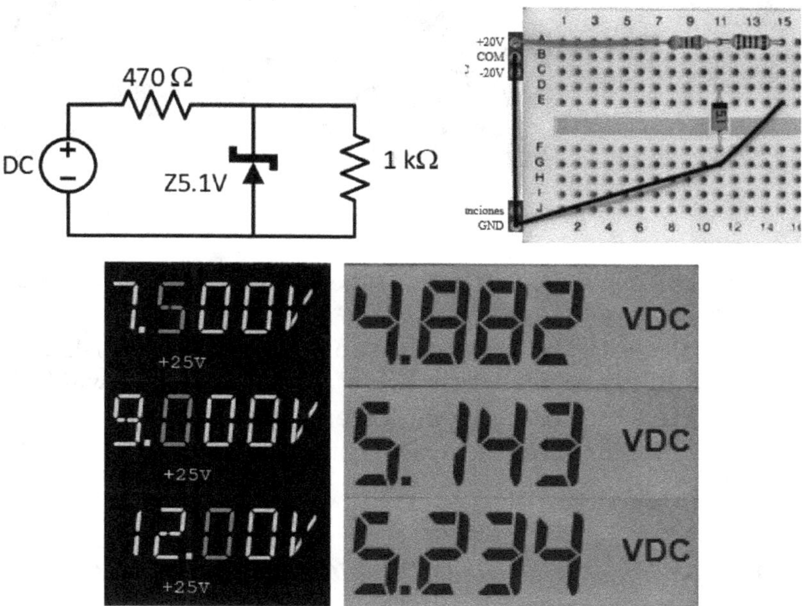

Figure 3.102. Zener diode circuit II.

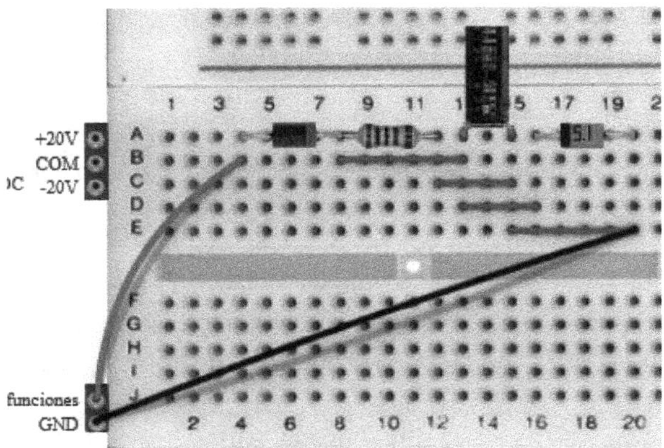

Figure 3.103. AC/DC converter with a Zener diode.

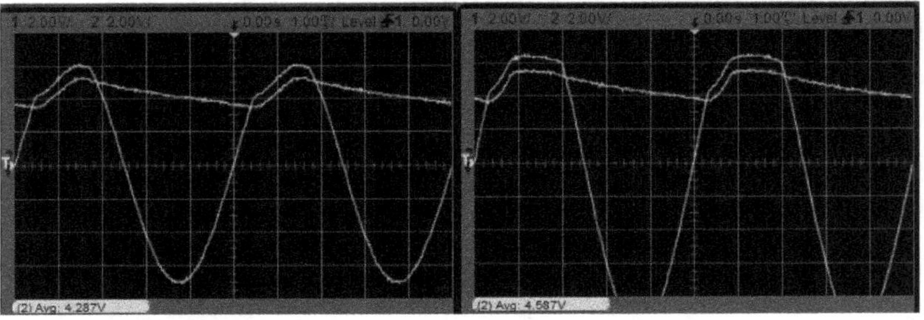

Figure 3.104. Output waveforms for an AC/DC converter with a Zener diode.

4. *Analysis and conclusion.* The Zener diode can be used to regulate the output voltage.

In the voltage divider, replace the two resistors with two 1-kΩ resistors and two 10-kΩ resistors. In this case, the output voltage will be half the input voltage (without using the Zener). What happens to the Zener diode? Does the output voltage remain constant? Why not for the 10-kΩ resistors?

3.4.10 *Experiment with Diode Circuit: Different Diodes*

1. *Introduction.* There are several types of diodes, depending on the semiconductor material used, which affects the voltage drop, among other characteristics. All of them have the same behavior but with some differences.

We will experiment with two different diodes to answer the following question: What happens when the diode changes from forward bias to reverse bias and vice versa? How much time is needed to change from one mode to the other one?

2. *Logical model.* When the diode is in direct polarization (forward bias), the current goes from the anode to the cathode, but if the polarization is changed, then the current will go in the opposite direction: from the cathode to the anode.

How much time do the carriers (electrons and holes) need to change their current direction?

From reverse bias to forward bias, only a few carriers need to change direction, but from forward bias to reverse bias, far more carriers need to change direction. Therefore, it seems that more time will be needed in the second situation.

3. *Experiment.* We set up the simple circuit of Figure 3.105 for two different diodes: 1N4007 and 1N4148.

If we excite the circuit with a square signal at 10 kHz, we can see how each diode behaves.

In Figure 3.106, the diode is the 1N4007. We can see that the diode needs more time to switch off than to switch on. The switch-off time is

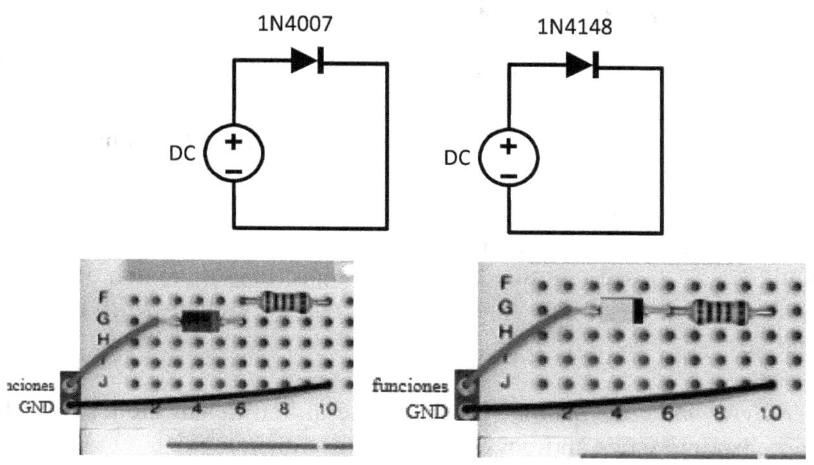

Figure 3.105. Diode circuits: 1N4007 and 1N4148.

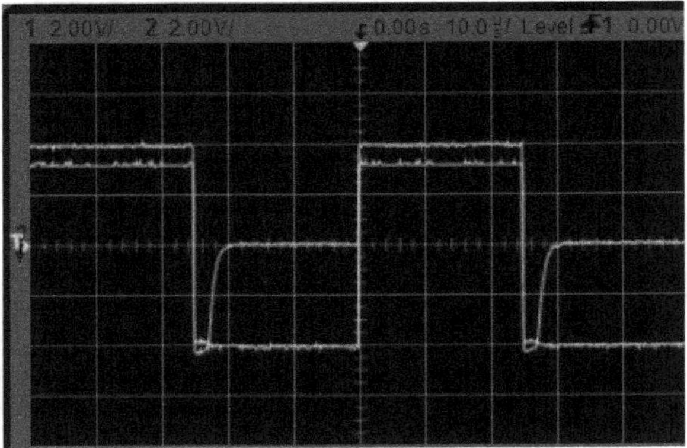

Figure 3.106. Output waveform for a 1N4007 diode.

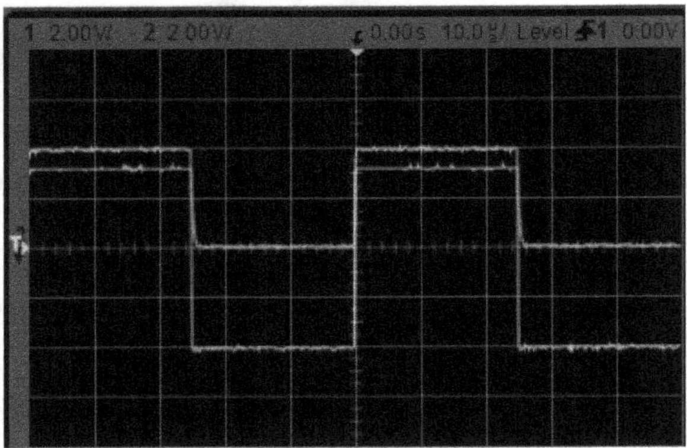

Figure 3.107. Output waveform for a 1N4148 diode.

more or less 4 μs (half of a 10-μs square). Even during this time, the voltage is negative, when it should be zero, and therefore, during this time, the current will be negative.

In Figure 3.107, we see the behavior of the 1N4148 diode. In this circuit, we again see that the diode needs more time to switch off than to switch on. In this case, however, the output voltage will never be negative.

4. *Analysis and conclusion.* There are several families of diodes with different characteristics and possibilities for designers.

The previous experiment is intended to show only one of them. The characteristics of a diode must be analyzed on the corresponding datasheet.

3.5 Other Circuits: Transistors and Operational Amplifiers

3.5.1 *Bipolar Transistor Analysis*

1. *Introduction.* The bipolar transistor is a three-terminal device that makes it possible to control a high-power output from a low-power input signal. The power to be regulated is applied between the collector (C) and emitter (E) terminals, and on the base terminal (B), the control signal is fed, enabling us to control power.

The bipolar transistor bases its operation on the control of the current flowing between the emitter and the collector, by means of the base current. In the VISIR, we can control this current to observe the variation in the transistor behavior. In an NPN transistor — the one that is available in the Deusto VISIR — the following equation applies:

$$I_E = I_B + I_C$$

2. *Logical model.* The operating point of the transistor, also known as the bias point, quiescent point, or Q-point, is defined by the steady-state DC voltage between the collector and the emitter (V_{CE}) and the current flowing through the collector (I_C). Biasing methods make the transistor work mainly in three regions:

- **Active region**: This is also known as the linear region. While in this region, a transistor acts as an amplifier. In this region, collector current is β times the base current.

$$I_C = \beta I_B$$

- **Saturation region**: Here, the transistor behaves as a closed switch, and has a shortening effect on its collector and emitter.

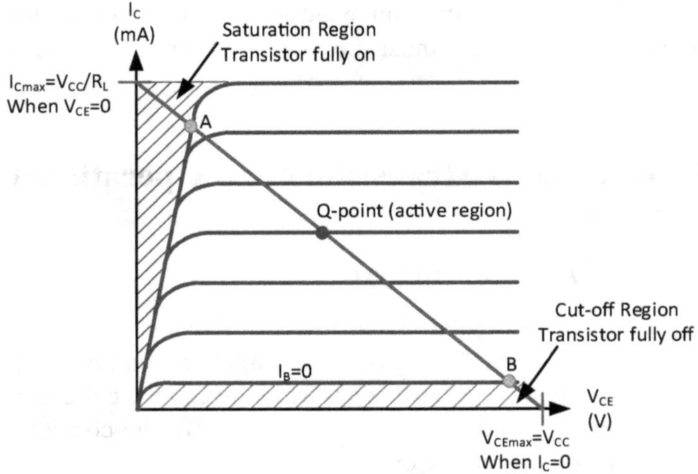

Figure 3.108. Operating regions of the transistor.

In this mode of operation, collector and emitter currents are maximum and both are forward-biased.

$$I_C = I_E$$

- **Cut-off region**: This is the region in which the transistor behaves as an open switch. The transistor has the effect of its collector and base being opened, with all the currents through the devices being zero.

From the characteristic curves of the transistor (Figure 3.108), and calculating the *Q*-point, the transistor will be biased and working in one of the three regions previously introduced.

3. *Experiment.* To perform this experiment, we will work with the circuit shown in Figure 3.109.

At the shared GitHub repository, the lab technician can download the BJT_circuit.max file.

The experiment consists of changing the emitter resistance and observing how the bias of the transistor changes. Thus, if $R_e = 220\ \Omega$, the *Q*-point will be in the active region, while if $R_e = 100\ \Omega$, the transistor will work in the saturation region.

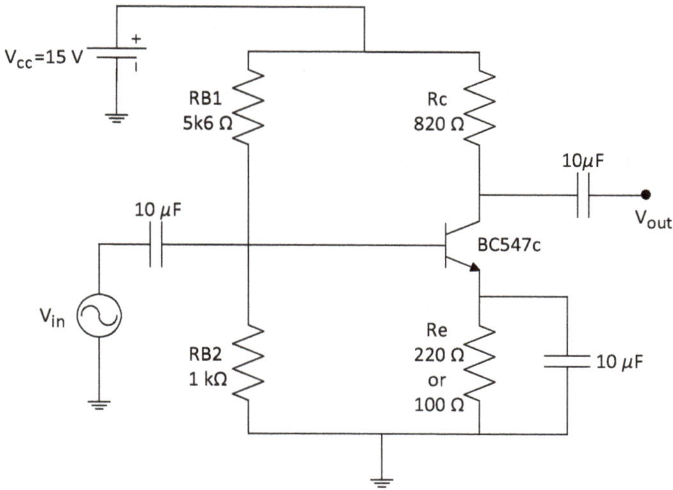

Figure 3.109. Bipolar transistor circuit.

Figure 3.110. DC circuit for active region analysis.

3.5.1.1 *Active region analysis*

To calculate the operating point of the transistor, we eliminate the input signal generator and leave only the supply voltage of the circuit, obtaining the circuit shown in Figure 3.110. For illustrative purposes, the emitter resistor, which for this experiment has a value of 220 Ω, is highlighted by a blue rectangle.

We will now measure the values indicated in Table 3.32 and check whether the transistor is indeed working in the active region.

Using these values, we can check that *Q*-point is placed at 7.52 V vs. 7.28 mA. If we represent this *Q*-point at the characteristic curve diagram, we obtain Figure 3.111 that confirms that the transistor is working in the active region.

The following checks can also be performed:

$$V_{CE} + V_{Rc} + V_{Re} = 7.52 \text{ v} + 1.618 \text{ V} + 5.782 \text{ V} = 14.92 \text{ V} \cong V_{CC} = 15 \text{ V}$$
$$V_{RB1} + V_{RB2} = 12.78 \text{ V} + 2.248 \text{ V} = 15.028 \text{ V} \cong V_{CC} = 15 \text{ V}$$

Table 3.32. Performed measurements to characterize the active region.

Measurement	Value
V_{CE}	7.52 V
V_{RE}	1.618 V
V_{RC}	5.782 V
V_{RB1}	12.78 V
V_{RB2}	2.248 V
I_C	7.28 mA
I_B	39 μA

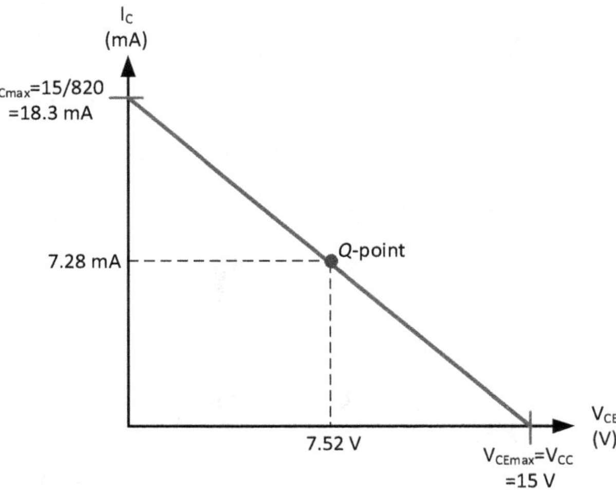

Figure 3.111. *Q*-point calculation at active region.

$$I_B = I_{RB1} - I_{RB2} = V_{RB1}/R_{B1} - V_{RB2}/R_{B2} = 12.78/5k6 - 2.248/1k$$
$$2.28\text{ mA} - 2.24\text{ mA} = 40\ \mu\text{A} \cong I_B = 39\ \mu\text{A}$$

Moreover, with these values, the β factor of the transistor can be calculated as follows:

$$\beta = \frac{I_C}{I_B} = \frac{7.28\,\text{mA}}{39\,\text{A}} \approx 182$$

Finally, one can observe the behavior of the circuit as an amplifier. If the function generator is set up with a 600-Hz and 100-mV$_{pp}$ signal, circuit gain can be calculated. The circuit plus the input and output signals displayed at the oscilloscope are shown in Figure 3.112.

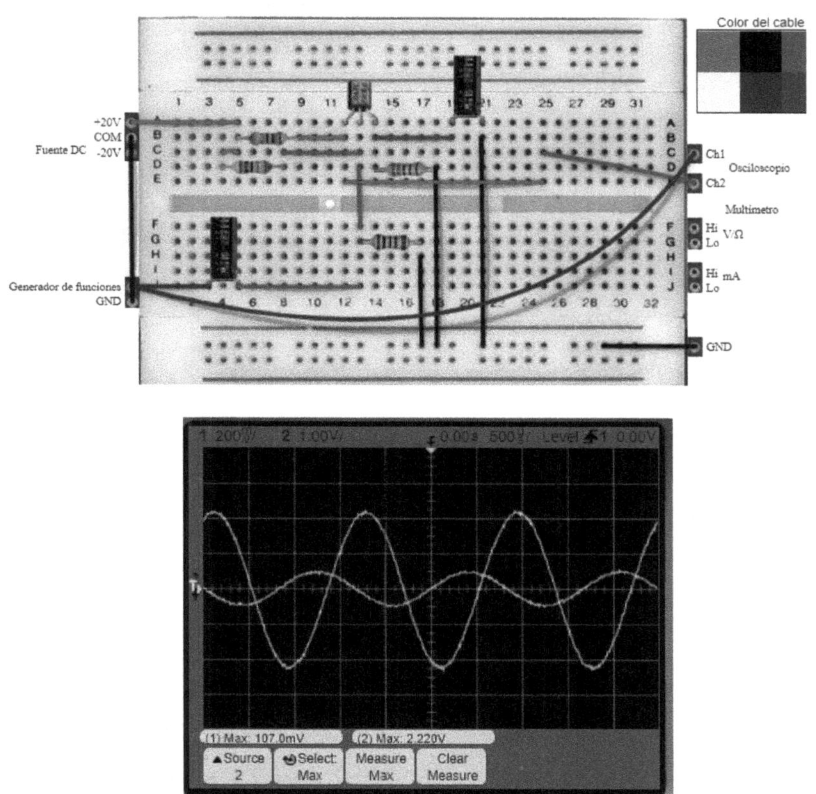

Figure 3.112. BJT transistor working as an amplifier at active region.

3.5.1.2 *Saturation region analysis*

To move the *Q*-point of the transistor to the saturation region, we only have to change Re resistor to 100 Ω. Then V_{CE} voltage drops to 1.116 V and the current through the collector increases to 15.13 mA. Measurements are included in Table 3.33, obtained via DC analysis of the circuit.

Thus, if we represent the Q-point using these values (Figure 3.113), we can check that the transistor will work in the saturation region when an input signal applies.

Table 3.33. Measurements taken to characterize the saturation region.

Measurement	Value
V_{CE}	1.116 V
V_{RE}	1.52 V
V_{RC}	12.41 V
V_{RB1}	12.82 V
V_{RB2}	2.20 V
I_C	15.13 mA
I_B	80 μA

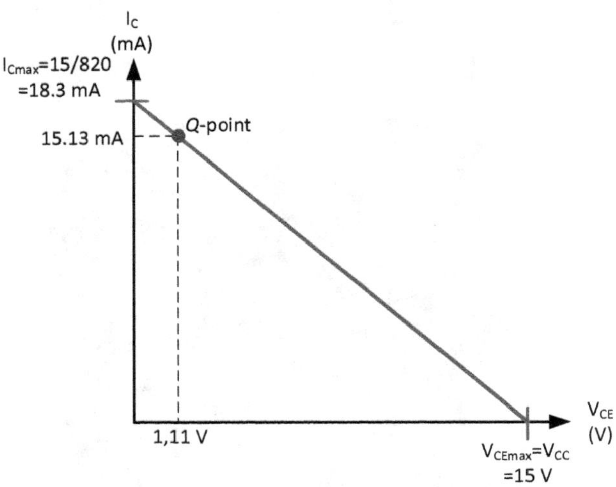

Figure 3.113. Q-point calculation at the saturation region.

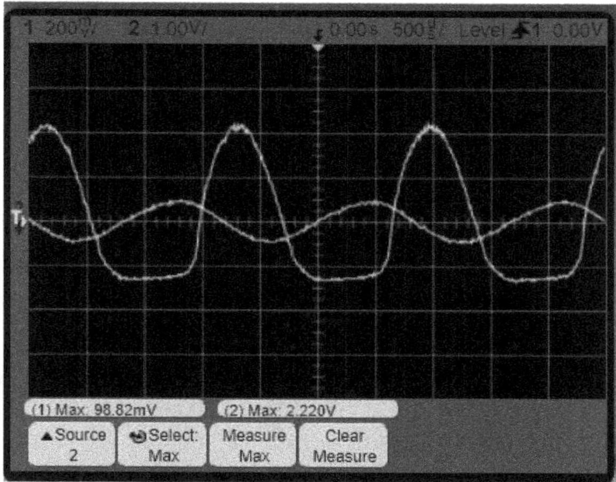

Figure 3.114. BJT transistor input vs. output signal working as amplifier in saturation region.

Now, if we feed the circuit, by setting the function generator with the same signal (600 Hz@100 mV), the effect of the saturation can be observed in Figure 3.114, where the negative cycle of the output signal is clipped.

Analysis and conclusion. By making minor changes in the circuit configuration (modifying the value of a resistor), we can test the behavior of a transistor in different operating zones. In addition, being a complex circuit, compared to the experiments in the previous sections, it allows us to review concepts such as Ohm's and Kirchhoff's laws, as well as to examine in detail the DC and AC analysis of a bipolar transistor.

3.5.2 *Operational Amplifier Circuits*

1. *Introduction.* An operational amplifier, or op-amp for short, is basically a device designed to amplify voltage (or current), which is commonly used with resistors and capacitors connected between their output and input terminals. Depending on the values and set-up of these feedback components, the function of the op-amp is determined.

It is a three-terminal device, two of which are high-impedance inputs. One of the inputs is called the Inverting Input, marked with a negative sign (–). The other input is called the Non-inverting Input, marked with a positive sign (+). The third terminal represents the output of the op-amp that can both sink and source either a voltage or a current.

2. *Logical model.* An operational amplifier consists of a very high-gain DC differential amplifier. Using one or more feedback circuits, designers can determine the response and characteristics of the resulting circuit. Basic functions or topologies using an op-amp are inverting, non-inverting, voltage follower, summing, differential, integrator, and differentiator amplifiers.

Thus, using VISIR, different applications can be configured using operational amplifiers. Given the particularities and the special nature of the remote laboratory, the real challenge for the laboratory technician and the teacher is to choose those circuits that contain common components and/or feedback networks so that the resources (relays of the switching matrix) of VISIR are optimized to the maximum.

3. *Experiment.* All the experiments proposed in the following can be configured by the laboratory technician using the .max files that can be found in the "1.5 Other circuits" directory at the GitHub repository.

The following experiments are intended to enable the student/user to analyze the behavior of operational amplifiers in various configurations, studying the relationship between the output and the input voltages.

As has been indicated, not all configurations have been included in the Deusto VISIR. In addition, since a potentiometer is not available, in order to provide the experiment with more versatility, we recommend the use of different feedback resistors to analyze different configurations of the same op-amp functionality.

The core of the proposed experiments is the most commonly available operational amplifier in basic electronic kits and projects, i.e., the μA-741 op-amp Integrated Circuit (IC). In all the set-ups, the IC must be powered by +15 VDC and –15 VDC voltages. In the following schematics, numbers indicate the pin terminals of the IC.

3.5.2.1 *Inverting amplifier*

When op-amp works in this set-up, the operational amplifier is connected with feedback to produce a closed-loop operation (Figure 3.115). It is important to recall that operational amplifiers obey two important rules: No current flows into the input terminals and the voltage at the negative terminal (V–) always equals the voltage at the positive terminal (V+). However, in a real experiment, such as those that we run in the VISIR, both rules are slightly broken.

Thus, the junction of the input and feedback branch is at the same potential as the positive terminal input (V+), which is at zero volts or ground, so the junction is a "Virtual Earth". Because of this virtual earth node, the input resistance of the amplifier is equal to the value of the input resistor, R_1, and the closed loop gain of the inverting amplifier can be set by the ratio of the two external resistors, following the schematic shown in Figure 3.116.

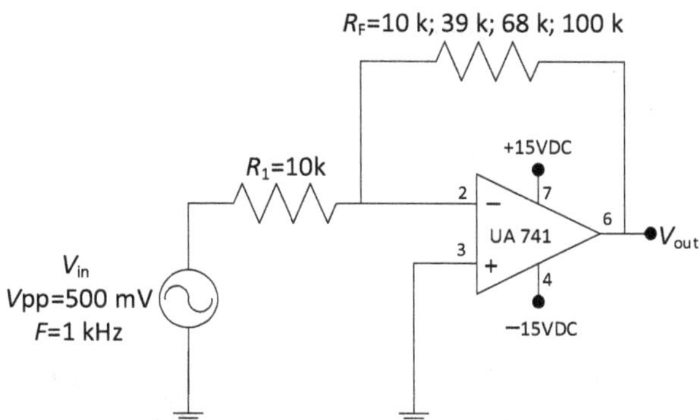

Figure 3.115. Op-amp operating as an inverter amplifier.

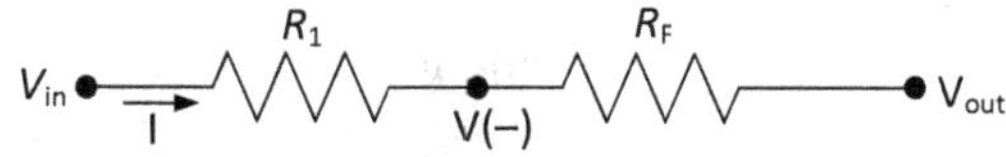

Figure 3.116. Equivalent circuit for inverting amplifier set-up.

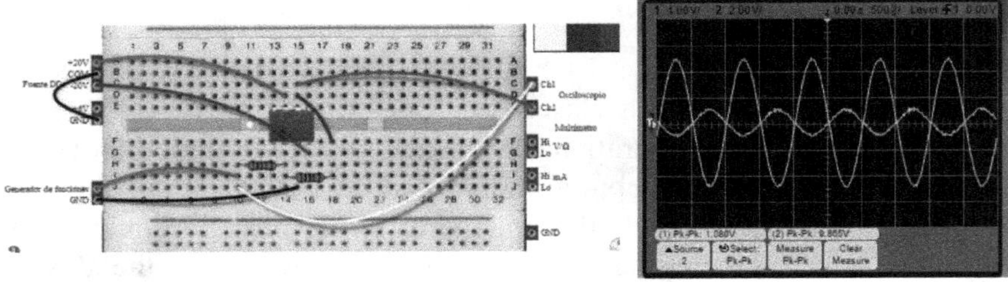

Figure 3.117 Implementation of op-amp operating as inverter amplifier, being R_F = 100 kΩ.

To solve this circuit, the following equations are proposed:

$$I = \frac{V_{\text{in}} - V(-)}{R_1} = \frac{V(-) - V_{\text{out}}}{R_{\text{F}}} \rightarrow \frac{V_{\text{out}}}{V_{\text{in}}} = -\frac{R_{\text{F}}}{R_1}$$

Observing the last equation, the negative sign equation indicates an inversion of the output signal with respect to the input, as it is 180° out of phase. This is due to the feedback being negative in value.

This effect can be analyzed thanks to the VISIR carrying out the circuit and measures shown in Figure 3.117. On the oscilloscope, it is clear that the signals are 180° out of phase, and comparing both peak-to-peak voltage measures, the ratio between both signals can be also calculated as R_F/R_1 = 100 k/10 k = 10.

3.5.2.2 *Non-inverting amplifier*

In this set-up (Figure 3.118), the function generator feeds the V_{in} terminal applied directly to the non-inverting input terminal. This means that the output gain of the amplifier is positive, in contrast to the "Inverting Amplifier" whose output gain is negative in value. The effect of this configuration is that the output signal is in phase with the input signal (Figure 3.119).

From the peak-to-peak measurements taken of both input and output signals, it can be seen first that both signals are in phase on the oscilloscope and that

$$\text{Gain} = \frac{R_1 + R_F}{R_1} = \frac{10\text{k} + 100\text{k}}{10\text{k}} = 11$$

$$V_{\text{out}} = \text{Gain} \times V_{\text{in}} \rightarrow 11.06(V_{\text{out\_pp}}) \cong 11 \times 1.08(V_{\text{in\_pp}})$$

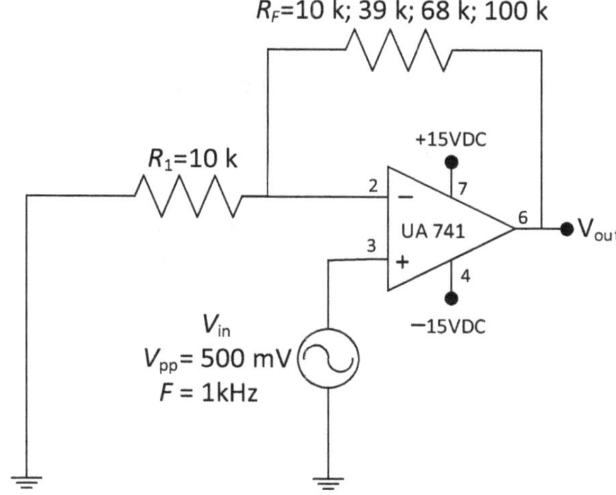

Figure 3.118. Op-amp operating as non-inverter amplifier.

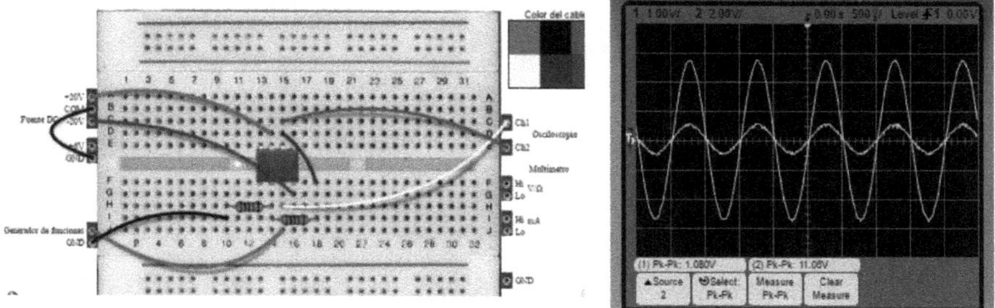

Figure 3.119. Implementation of op-amp operating as non-inverter amplifier, being $R_F = 100\ \text{k}\Omega$.

3.5.2.3 *Differential amplifier*

If the basic theory about the op-amp is analyzed in depth (which is not within the scope of this book), all op-amps work as differential amplifiers because of their input configuration. However, if different voltage sources are connected to both input terminals [V (+) and V (–)], the output voltage will be proportional to the difference between both input voltage signals.

As there are two function generators available in the VISIR, we propose a layout in which the voltage applied to the positive input terminal can be either ground or the same input signal applied to the negative terminal (dot lines in Figure 3.120). With either configuration, the output signal is proportional to the difference in voltages applied to the two inputs.

If the circuit is analyzed applying basic op-amp theory and Ohm's and Kirchhoff's laws, the next reasoning is obtained as follows:

$$I_1 = \frac{V_1 - V_a}{R_1}; \quad I_2 = \frac{V_2 - V_b}{R_2}; \quad I_F = \frac{V_a - V_{\text{out}}}{R_3}$$

In an op-amp → $V_a = V_b$

Voltage divider at $V(+)$ terminal → $V_b = V_2(R_4/(R_2 + R_4))$

$$\textit{If } V_2 = 0, \quad \text{then } V_{\text{out}(a)} = -V_1(R_3/R_1)$$

$$\textit{If } V_1 = 0, \text{ then } V_{\text{out}}(_b) = -V_2(R_4/(R_2 + R_4))((R_1 + R_3)/R_1)$$

$$V_{\text{out}} = V_{\text{out}(a)} + V_{\text{out(b)}}$$

$$\text{V}_{\text{out}} = -\text{V}_1(\text{R}_3/\text{R}_1) + V_2(R_4/(R_2 + R_4))((R_1 + R_3)/R_1))$$

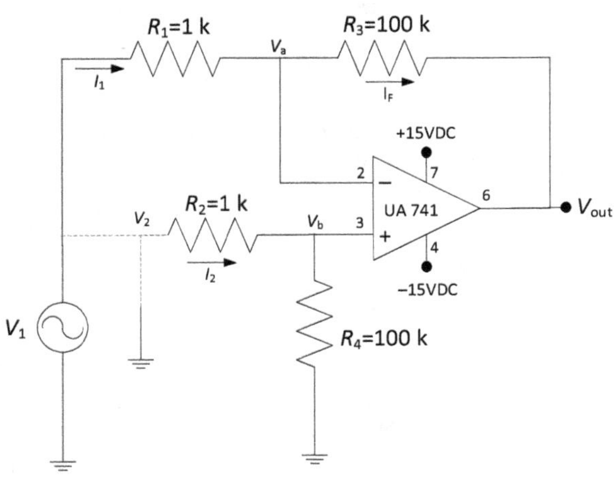

Figure 3.120. Op-amp operating as a differential amplifier.

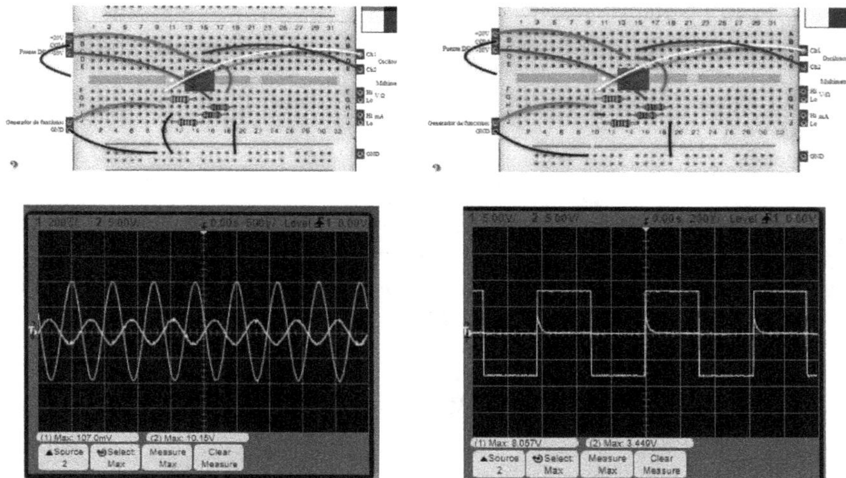

Figure 3.121. Implementation of an op-amp operating as non-inverter amplifier. On the left, circuit and scope with the R_1 resistor grounded, and on the right, $V_1 = V_2$.

If $R_1 = R_2$ and $R_3 = R_4$, the above transfer function for the differential amplifier can be simplified to the following formula:

$$V_{\text{out}} = \frac{R_3}{R_1}(V_2 - V_1)$$

Experiments performed in VISIR are shown in Figure 3.121, setting up the function generator to provide a squared signal V_1 = 1600 Hz@8 V_{pp}.

Analysis and conclusion. The VISIR remote laboratory allows for the implementation of complex circuits, using components, such as transistors or operational amplifiers. The main limitation is the maximum number of nodes a circuit can have (from node A to node I plus ground). It is the joint task of the lab technician and the teacher to agree on the circuits to be made available to users so that the maximum number of components can be reused, and the switching matrix can be optimized.

In the case of circuits with operational amplifiers, this optimization implies that it is not possible to measure the currents in all branches of the circuit, limiting the measurements to voltages with the multimeter (and the underlying possibility of theoretically calculating the currents) and the oscilloscope.

Reference

Swartling, M., Bartůněk, J. S., Nilsson, K., Gustavsson, I., & Fiedler, M. (2012). Simulations of the VISIR Open Lab Platform. In *2012 9th International Conference on Remote Engineering and Virtual Instrumentation (REV)* (pp. 1–5). doi: 10.1109/REV.2012.6293108.

Part 3

Research and Reflections on VISIR

Chapter 4

VISIR Around the World in a Nutshell

4.1 Introduction

This chapter aims to provide a general view of the creation of the VISIR remote laboratory and its extension to the entire world, with a presence in every habitable continent, i.e., America (North–Central–South), Africa, Asia, Europe, and Oceania.

4.2 A Brief History of VISIR and Its Creator: Ingvar Gustavsson

Prof. Ingvar Gustavsson started developing an initial version of the remote laboratory called Virtual Instrument System In Reality (VISIR), in 1999, at the Blekinge Institute of Technology (BTH), in Sweden. His goal was to allow students freely to perform available experiments while learning about electric and electronic circuits. This is in line with his philosophy of experimenting, which "*... required [users] to formulate a useful 'question' to Nature and to interpret its answer i.e., to be fluent in the language of nature*", and paraphrases Max Planck's definition of experiments as a question posed to nature.

Ingvar's first publication (in English) about the possibility of performing remote experiments with electrical circuits dates back to 2001 (Gustavsson, 2001). In that paper, Ingvar acknowledges "*the support of* [...] *the DAL/MAL project with funding from DISTUM, the Swedish Agency for Distance Education*". These two projects (DAL/MAL) are described in an earlier paper by Håstad, Lundkvist, and Wikström (2001), which provides details on how Ingvar might have become involved in the project:

- BTH was involved in the DAL project, which included the development of distance learning materials for analog electronics.
- "*In the DAL project, laboratory work was identified as a major obstacle when trying to arrange distance education in the natural sciences and technical subjects*".
- The subsequent MAL project worked with those subjects, including circuit theory (which includes analog electronics), each subject leading to a specific subproject.
- "*The objective of the [...] subproject is [...] to improve access to laboratory experiments by making remote experimentation accessible by use of the Internet. Anyone can now do experiments in Circuit Theory from anywhere using a client PC connected to a laboratory server at BTH. Further information and client software can be downloaded from http://www.its.bth.se/courses/eta014/distanslabbar/english*".

The link is no longer active. However, using the Internet Archive (2021), it is possible to access a snapshot of that same page, taken on November 11, 2004, as illustrated in Figure 4.1.

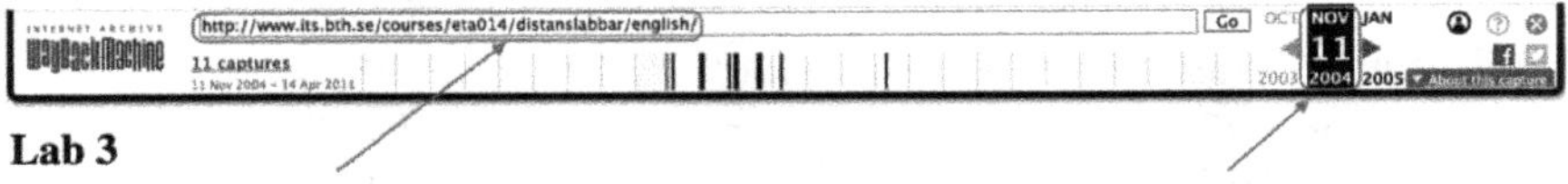

Lab 3

This laboratory involves real experiments arranged so that they may be carried out remotely via the Internet. The laboratory is part of an effort to incorporate distance learning.

Two compatible arrangements exist. The first (port 6341) contains conventional instruments controlled by the operator through a GPIB instrument bus, which has a fairly low transfer speed. The second (port 6342) consists of PXI instruments in the form of circuit cards, which are connected to the PC's PCI-bus via optical fibres. These instruments do not have normal front panels. The panels are virtual and are shown on the PC screen. The operator adjusts the settings by turning knobs and controls with the mouse. The five circuits in the laboratory instructions have been wired together on prototype cards, one for each arrangement. Sources and meters are connected to the circuits via switch units, one for each arrangement.

The laboratory involves no physical connections to be made by the student, however a certain understanding of handling instruments such as oscilloscopes, function generators, multi-meters and power supplies is necessary.

The experiments are described in the laboratory instructions. To carry out the experiments remotely requires a client, which works in Windows 9x, NT and 2000. Begin by downloading. Before installation you must uninstall the old client if you have one. If you have some ocx files that is too old the installer wants to replace them and this requires write access to the system directory.

The instructions are written for someone carrying out the laboratory in the traditional manner. The distance instructions add on describes how to carry out the laboratory remotely.

The server address is now 194.47.132.148. Some days during August and September the server will not be available. Please send a mail to ingvar.gustavsson@bth.se if you want use it .

The server connected to port 6341 is not in use.

Figure 4.1. Snapshot of the web page accessible through link http://www.its.bth.se/courses/eta014/distanslabbar/english, taken on November 11, 2004. Retrieved from http://web.archive.org/web/20041111175047/http://www.its.bth.se/courses/eta014/distanslabbar/english/.

Figure 4.1, corresponding to a webpage accessible through the link included in the paper written by Håstad, Lundkvist, and Wikström (2001), about the DAL/MAL projects, confirms the involvement of Ingvar Gustavsson, who continued to publish about his work, alone (Gustavsson, 2002a, 2002b, 2003a, 2003b, 2003c, 2003d, 2003e), until 2004. In that year, Ingvar published four papers, three of them with co-authors, which included a new link to the remote laboratory, still active today (July 2021), i.e., http://distanslabserver.its.bth.se/. An interesting detail about 2004 is that it corresponds to the year a young software engineer, named Johan Zackrisson, started working with Ingvar Gustavsson.

The two following years (2005 and 2006) were not so prolific in bibliographical terms, with just two publications per year, yet it was in February 2006 that the expression Virtual Instrument Systems in Reality (VISIR) first appeared in a publication. Quoting from Gustavsson *et al.* (2006):

> A consortium chaired by BTH will be formed to expand and disseminate our unique electronics laboratory internationally within the framework of a proposed project known as **VISIR** (**Virtual Instrument Systems in Reality**). Universities and other organizations will be invited to participate in the development and dissemination together with BTH as part of the VISIR project.

The VISIR project first gained public recognition following an interview with Ingvar Gustavsson, published on October 30, 2006, in *Sydöstran*, a daily newspaper in Karlskrona, Sweden, covering local news, sports, business, politics, and community events. The newspaper interview (in Swedish) is reprinted in Figure 4.2. An interesting aspect mentioned in this interview is the invitation extended to Ingvar Gustavsson to participate in an International Meeting on Professional Remote Laboratories, organized by Javier García Zubía (!), at the University of Deusto, Bilbao, Spain, on November 16–17, i.e., a fortnight after the interview. Quoting from Figure 4.2 (translated from Swedish, using Google translator):

> In two weeks, he will be in Bilbao, in Spain, where the ten universities that have come the furthest with distance laboratories will meet to compare their projects and discuss cooperation.

SYDÖSTRAN | MÅNDAG DEN 30 OKTOBER 2006 RONNEBY

Prisat distanslabb sprider sina idéer över Europa

ÖPPNAR KÄLLKODEN// Vill bygga nätverk

RONNEBY

Alla sladdar är kopplade via datorn, och när man rör musen mot knappen på oscillatorbilden på skärmen rör knappen på sig och det börjar hända saker i en riktig oscillator på BTH i Ronneby.

– Det här är ingen datorsimulering, utan man gör experimentet på riktigt med riktiga instrument, fast via datorn, och man kan sitta var som helst i världen och göra det, berättar Ingvar Gustavsson som byggt upp distanslaboratoriet på Blekinge tekniska högskola.

Nu har han fått högskolans pedagogiska pris för sina insatser.

Distanslabbet används i utbildningen Ingenjör on line, en distansutbildning i samarbete mellan åtta olika högskolor. De blivande ingenjörerna kan sitta runt om i landet vid sina datorer och göra sina elektronikexperiment på distanslabbet i Ronneby.

Nyckel till labbsalen

– Men det är också mycket använt av våra egna campusstudenter, berättar Ingvar Gustavsson.

Laborationssalarna med sina dyrbara instrument står för det mesta låsta.

Men genom distanslaboratoriet får studenterna så att säga en egen nyckel till labbet och kan själva sitt hemma och gå igenom laborationerna så många gånger de behöver.

De experiment man kan göra i distanslabbet i Ronneby är inom ellära, men man håller också på att bygga upp en experimentutrustning i vibrationsanalys.

Själva labbet ryms på och under ett skrivbord inne i en skrubb, längst in bland en massa andra apparater. Vibrationsmätningsutrustningen hänger i en hemsnickrad träbox.

– Det är minst sagt utrymmessnålt, konstaterar Ingvar Gustavsson.

Små resurser

Distanslabbethar byggts upp med små resurser och länge i ekonomisk motvind. Mycket av programmeringsarbetet gjordes av sommarjobbande studenter.

– Men nu har det lossnat. BTH har beslutat att göra en satsning på det här, och jag också fått pengar från Vinnova. Så just nu är jag ganska välfinansierad, säger Ingvar Gustavsson.

– Det passar väl i tiden, eftersom det är mycket på gång på det här området ute i Europa.

Till Spanien

Om två veckor ska han Bilbao i Spanien där det tiotal universitet som kommit längst med distanslaboratorier ska träffas för att jämföra sina projekt och diskutera samarbete.

– Vi har beslutat att släppa vår källkod. Den kommer att vara tillgänglig för alla som vill bygga upp en liknande verksamhet. Men det blir ett centrum här på BTH där vi håller programvaran uppdaterad.

– Vi skulle vilja bygga upp ett nätverk där man kan använda experiment från andra som man inte har själv. Man kopplar upp sig mellan servrarna utan att studenterna märker någon skillnad.

– Det är sådana saker vi ska diskutera i Spanien. Och till nästa år har vi tänkt bjuda in dem till oss i Blekinge, säger Ingvar Gustavsson.

JAN HINDERSON

PÅ DISTANS. Ingvar Gustavsson styr oscillatorn i distanslabbet direkt via musen och dataskärmen. FOTO JAN HINDERSON

LABB. Hela uppsättningen av traditionella laborationsinstrument...

DISTANS. ... flyttar in i datorn, som kan stå var som helst i världen.

Figure 4.2. Interview with Ingvar Gustavsson, Sydöstran newspaper, October 30, 2006.

> We have decided to **release every key code**. It will be available to anyone who wants to build a similar lab. But there will be a center at BTH where the software will be updated.
> We would like to **build a network where you can use experiments from others** that you do not have yourself. You connect between the servers without the students noticing any difference.
> These are the things we are going to discuss in Spain. And next year we intend to invite them to visit us in Blekinge, says Ingvar Gustavsson.

These two objectives (open/reuse software plus share experiments) formed the basis for expanding VISIR from Sweden to the whole world.

4.2.1 *Ingvar Gustavsson's Short Biography*

Ingvar Gustavsson (15 October 1943–2 October 2017) was born in Karlskrona, Sweden. He received the M.S.E.E. and Dr.Sc. degrees in electrical engineering from the Royal Institute of Technology (KTH), Stockholm, in 1967 and 1974, respectively. After completing his military service in 1968, he worked as a Development Engineer at Jungner Instrument AB in Stockholm. In 1970, he joined the computer vision project SYDAT at the Instrumentation Laboratory, KTH. In 1982, he was appointed head of the Instrumentation Laboratory. Together with another research scientist, in 1983, he founded a private company providing automatic inspection systems for industrial customers. In 1994, he returned to the academic world to take up a position as Associate Professor of Electronics and Measurement Technology at BTH, Sweden. In 1999, he started a remote laboratory project at BTH known today as VISIR. He partially retired from office in 2012 in order to concentrate on activities related to VISIR.

His research interests are the areas of instrumentation, remote laboratories, industrial electronics, and distance learning. Dr. Gustavsson has resigned membership of international committees, but he continued as a member of the IEEE and of Swedish professional societies until his death.

Both VISIR and Dr. Gustavsson received international recognition. VISIR was presented with the GOLC Online Laboratory Award 2015 in the category "Remote Controlled Laboratory", and Dr. Gustavsson received (post-mortem) the SEFI Francesco Maffioli Award of Excellence for Developing Learning and Teaching in Engineering Education, in 2018.

Both awards were given in the first year they were created, i.e., the Global Online Laboratory Consortium (GOLC) created the GOLC Online Laboratory Award in 2015 (GOLC, 2021) and the European Society for Engineering Education (SEFI) created the SEFI Francesco Maffioli Award in 2018 (SEFI, 2018).

4.3 From Sweden to the Whole World

2007 was a turning point for VISIR. The first installation outside Sweden was at the University of Applied Sciences FH Campus Wien, Austria, where Thomas Fischer was the local contact. Figure 4.3 illustrates Thomas Fischer's personal memoir of this event.

Another two important milestones achieved in 2007 were the completion of the first Master Thesis on VISIR (Rosén and Nilsson, 2007) and the first publication about VISIR authored by a person outside Ingvar's group at BTH (Fischer, 2007).

In the following years, six new VISIR nodes were installed outside Sweden, at the University of Deusto (UDeusto), Bilbao, Spain, in 2008; Carinthia University of Applied Sciences (CUAS), Austria, in 2009; Polytechnic of Porto — School of Engineering (IPP/ISEP), Portugal, in

Figure 4.3. VISIR installation at FH Campus Wien, Austria.

2010; National Distance Education University (UNED), Madrid, Spain, in 2011; Indian Institute of Technology Madras (IIT-Madras), India, in 2012; and, finally, College of the North Atlantic, Qatar (CAN-Qatar), in 2012.

In the period 2011–2012, there was intensive activity around VISIR. A Special Interest Group (SIG) was set up under the umbrella of the International Association for Online Engineering (IAOE), which subsequently became the VISIR Federation, as illustrated in Figure 4.4.

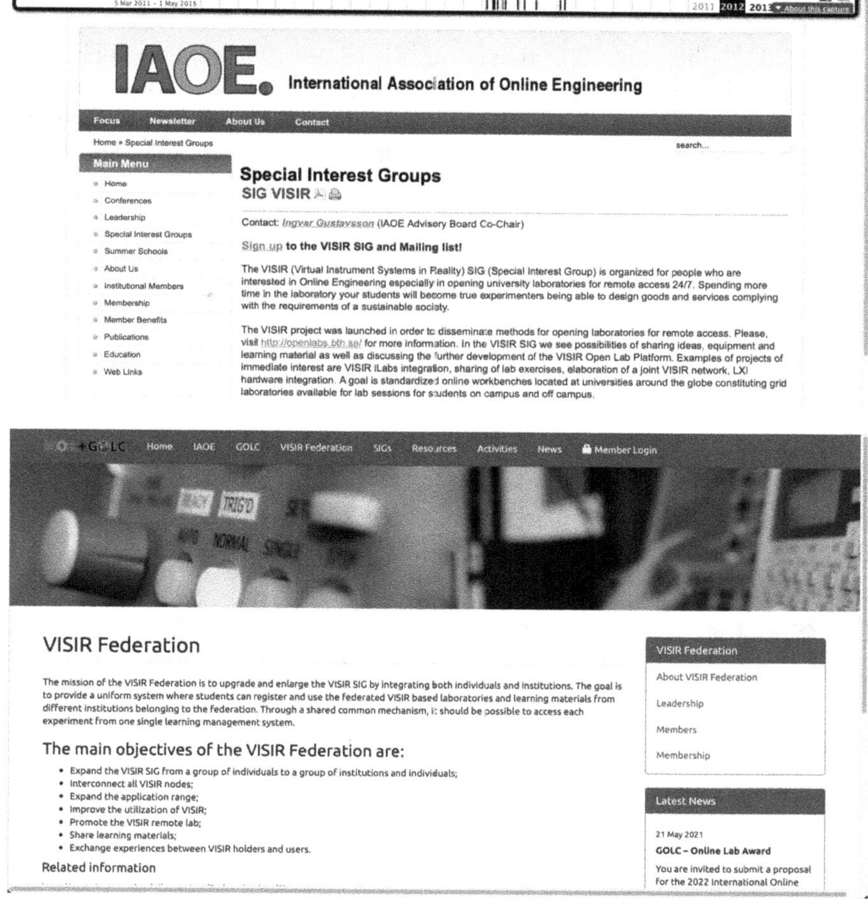

Figure 4.4. The VISIR SIG (top — snapshot taken on 2012.01.09) and the VISIR Federation (bottom — snapshot taken on 2021.07.23) webpages hosted by IAOE.

At the same time, Ingvar Gustavsson encouraged a group to prepare a proposal for the FP7-ICT-2011-8 Call, targeting in specific objective ICT 2011.8.1 Technology-Enhanced Learning. The proposal, named Innovative Technology-Enhanced Laboratory Learning Infrastructure (iTELLI), had VISIR as its core remote laboratory to extend access to experiments with basic electrical circuits to thousands of students in Europe. Although iTELLIE was not selected, the winning proposal, named Go-Lab, included two consortium members that offered VISIR as part of the remote experiments made available through the Go-Lab project. Figure 4.5 presents a snapshot of the Go-Lab portal, displaying remote experiments based on VISIR.

In 2012, the first PhD thesis on VISIR written outside BTH was publicly defended by Unai Hernández-Jayo, at the University of Deusto, Spain (Hernández-Jayo, 2012). Although this will be referred to again in the following section, which presents a survey on bibliography and PhD/MSc theses on VISIR, the important aspect here is that the solution developed by Hernández-Jayo (2012) was later implemented in another Higher Education Institution (HEI), in 2014, i.e., at the Batumi Shota Rustaveli State University, in Georgia, under the framework of the ICo-op project (2012).

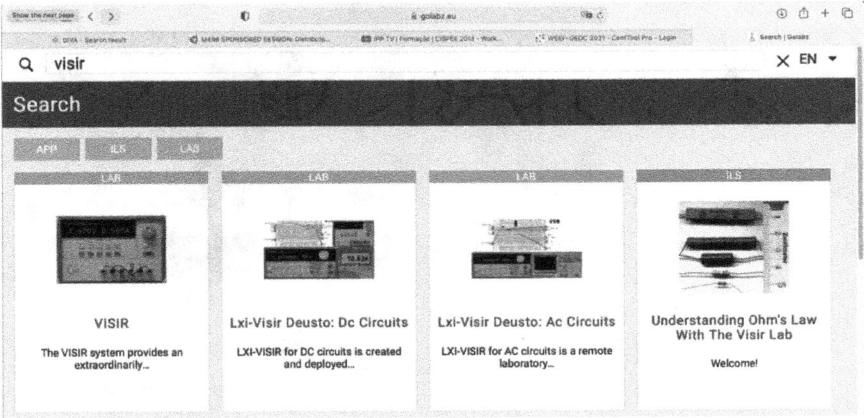

Figure 4.5. Remote experiments based on VISIR and delivered through the Go-Lab portal (snapshot taken on 2021.07.23).

In the following year, 2015, a project proposal to install five new VISIR nodes in Argentina and Brazil, under the framework of the ERASMUS+ program, was submitted and approved. The project, named "Educational Modules for Electric and Electronic Circuits Theory and Practice following an Enquiry-based Teaching and Learning Methodology supported by VISIR" or VISIR+ in short, attracted European partners who already had VISIR nodes installed on their premises (BTH, IPP/ISEP, CUAS, UNED, and UDeusto), and the Latin-American partners that would receive the new VISIR nodes (UFSC, PUC-Rio, IFSC, UNR, and UNSE). The VISIR+ project (2015) lasted for 2½ years, i.e., from October 2015 until April 2018 and it brought, in that period, the best and worst news to a growing VISIR Community of Practice (CoP).

An unprecedented number of publications, invited keynote speakers, and students and teachers using VISIR; a new European project for the creation of the first federation of VISIR remote labs, i.e., the "Platform Integration of Laboratories based on the Architecture of visiR" (PILAR) project (2016); three PhD theses with a strong focus on the VISIR remote laboratory (Arguedas-Matarrita, 2017; Najimaldeen, 2017; García-Loro, 2018); and the foundation of a start-up company named LabsLand, which would install three new VISIR nodes at the National University of Distance Education[1] (Costa Rica), the Technical University of Dortmund[2] (Germany), and the University of Georgia[3] (USA), were some of the positive stories shared among the VISIR SIG in that period. The saddest news would come on October 2, 2017, with the passing of Ingvar Gustavsson, the creator of VISIR.

Celebrating Ingvar's contribution to such a large community of users, the VISIR SIG submitted a proposal to the Francesco Maffioli Award for Excellence in Engineering Education, created by the European Society for Engineering Education in 2018 (SEFI, 2018). The prize was rightfully

---

[1] https://labsland.com/blog/en/2018/07/18/electronics-laboratory-deployed-in-costa-ricas-distance-university-uned/.

[2] https://labsland.com/blog/en/2019/03/06/new-electronics-laboratory-at-tu-dortmund/.

[3] https://labsland.com/blog/en/2019/04/18/electronics-laboratory-deployment-at-uga/.

awarded to Ingvar Gustavsson, in September 2018. In the Jury's own words:

> Prof. Ingvar Gustavsson won the award for his contribution in using remote experiments using Virtual Instrument System In Reality (VISIR). He started developing the remote lab in 1999, at the Blekinge Institute of Technology (BTH). His goal was to allow students to freely perform available experiments, while learning about electric and electronic circuits. Prof. Gustavsson's vision enables students to freely perform experiments, either locally or remotely.

To conclude this section, and summing up until the present day (September 2022), VISIR has been installed in 14 countries, including Argentina, Australia, Austria, Brazil, Costa Rica, Georgia, Germany, India, Morocco, Portugal, Qatar, Spain, Sweden, and the United States of America. The following list identifies the existing VISIR nodes, stating the host institution, city, country, and year of installation. Next, Figure 4.6 provides a world map view with the existing VISIR nodes and, finally, Figure 4.7 illustrates some moments associated with the installation and/or usage of several VISIR nodes.

1. Blekinge Institute of Technology (BTH), Karlskrona, Sweden, 1999.
2. FH Campus Wien University of Applied Sciences (FH-Wien), Vienna, Austria, 2007.
3. University of Deusto (UDeusto), Bilbao, Spain, 2008.
4. Carinthia University of Applied Sciences (CUAS), Villach, Austria, 2009.
5. Polytechnic of Porto — School of Engineering (IPP/ISEP), Porto, Portugal, 2010.
6. National University for Distance Education (UNED), Madrid, Spain, 2011.
7. Indian Institute of Technology Madras (IIT-Madras), India, 2012.
8. College of the North Atlantic (CAN-Qatar), Doha, Qatar, 2012.
9. Batumi Shota Rustaveli State University (BSU), Batumi, Georgia, 2014.

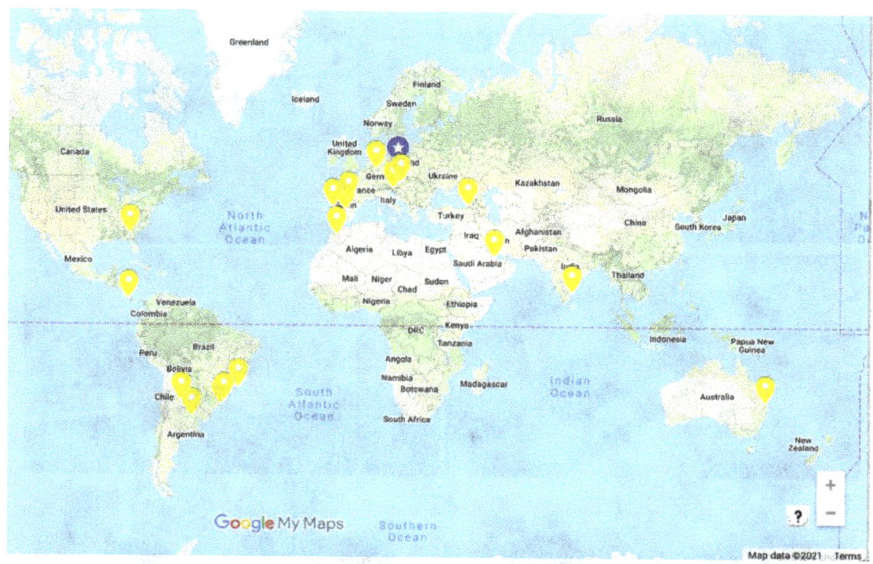

Figure 4.6. VISIR nodes installed around the world (2022).

10. Pontifical Catholic University of Rio de Janeiro (PUC-Rio), Rio de Janeiro, RJ, Brazil, 2016.
11. Federal University of Santa Catarina (UFSC), Araranguá, SC, Brazil, 2017.
12. Federal Institute of Santa Catarina (IFSC), Florianopolis, SC, Brazil, 2017.
13. National University of Rosario (UNR), Rosario, Argentina, 2017.
14. National University of Santiago del Estero (UNSE), Santiago del Estero, Argentina, 2017.
15. University of Hassan 1st (UH1), Settat, Morocco, 2018.
16. Universidad Nacional de Educación a Distancia (UNED), Costa Rica, 2018.
17. Technical University of Dortmund (TUDo), Germany, 2019.
18. University of Georgia (UGA), Athens, GA, USA, 2019.
19. University of Southern Queensland (USQ), Toowoomba, SQ, Australia, 2020.

Figure 4.7. VISIR installations/workshops/project meetings/interviews, etc.

4.4 A Survey on Bibliography and PhD/MSc Theses on VISIR

Silva *et al.* (2021) have produced a VISIR bibliographical reference list, which compiles close to 230 entries or publications. The authors clarify that to be included in this bibliographical reference, VISIR had to be either the main or a major subject in the referred publication. There are additional publications that refer to or mention VISIR that were not included due to the minor relevance VISIR had in the overall publication. As a result of this effort, a database with the complete VISIR bibliographic reference was built, using Mendeley Desktop, and made available as a BibTeX file at http://www2.isep.ipp.pt/cieti/uploads/VISIR_2001_2021.bib. The authors further refer the database which will be regularly updated and maintained by the VISIR federation (IAOE, 2021), so any researcher interested in VISIR may start from a complete reference list.

The present list version dates from 2001, with the first tracked publication by Gustavsson (2001), and includes all publications until August 2022. It includes Book Chapters (BC), Conference Proceedings (CP), Journal Articles (JA), Keynotes (KN), Magazine Articles (MA), Posters (PST), Reports (RPT), and Theses (TH), as depicted in Table 4.1. Figure 4.8 illustrates the number of publications per year (a) and the number of publications per type (b), i.e., BC, CP, JA, KN, MA, PST, RPT, and TH.

Table 4.1. Publications about VISIR from 2001 until August 2022.

Type	Year																						Total
	'01	'02	'03	'04	'05	'06	'07	'08	'09	'10	'11	'12	'13	'14	'15	'16	'17	'18	'19	'20	'21	'22	
BC			1	1			1				2	1	2			1	4	3	1	1	1		19
CP	1	2	3	2	2	3	3	3	4	6	11	11	6	15	5	8	22	20	11	7	14	4	163
JA			1			1		2	2	1	4	4	2	4	5	3	2	3	4	2	4	3	47
KN										1				1		1	2		1				6
MA										1													1
PST				1								1		1									3
RPT											1		1	1	1	2					1		7
TH							1			1		1	1	2	1		2	2		1	1	2	15
Total	1	2	5	4	2	4	5	5	6	10	18	18	12	24	12	15	32	28	17	11	21	9	261

Notes: Legend: BC (Book Chapter); CP (Conference Proceedings); JA (Journal Article); KN (Keynote); MA (Magazine Article); PST (Poster); RPT (Report); and TH (Thesis).

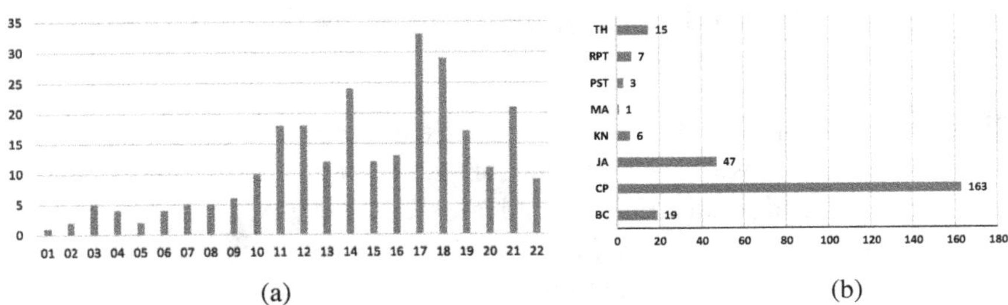

Figure 4.8. (a) Publications per year about VISIR and (b) publications per type about VISIR.

Figure 4.8(a) evidences the most prolific period (2017–2018) when the VISIR+ and PILAR projects were both active. Figure 4.8(b) shows that nearly 90% of the publications about VISIR correspond to conference proceedings, journal articles, or book chapters. Of these three publication types, the majority correspond to conference proceedings, with over 62% (i.e. 163/261) of all publication types. In addition to these more simplistic overviews, Silva *et al.* (2021) also present a series of graphs depicting:

- The evolution of authorships, authors, and publications, which also evidences the same peak in the period 2017–2018. An interesting number, depicted in the graph, is the number of different authors who have published work on VISIR, which reaches a maximum of nearly 80 individuals, in 2018. This number evidences a strong and active international community around the VISIR remote lab.
- The distribution of authors vs. their 20-year publication rate (i.e., calculated as the total number of publications over the total number of publishing years for each author). The graph shows that more than 100 authors publish, on average, one publication per year while, at the opposite end, one author has been able to publish, on average, nine publications per year.
- The production and collaboration rates over the 20-year period. Production rate is referred to as the number of authorships per author, while collaboration rate is the number of authors per publication. The graph also represents the change in these two indicators over the 20-year period. The average number of authors per publication is five, reaching a maximum of 9.5 authors per

publication in 2018. The average of authorships per author is slightly above two, with the maximum in 2003, corresponding to the five publications single-authored that year by Ingvar Gustavsson.

- The representation of the number of authors vs. number of one, two, three, or more publications per year, over the 20-year period. The graph exhibits several peaks for each class (one, two, three, or more), the latest (three or more publications) situated in 2018.
- The representation of the number of publications vs. number of authors, over the 20-year period. The extremes are 34 single-authored publications and one publication with 34 authors (Alves *et al.*, 2016).
- The number of publications per country (based on the first author's institution). The top 3 countries are Spain, Sweden, and Portugal, with 60, 52, and 45 publications, respectively.
- The top 10 fora, with a clearly preferred conference, that is, the International Conference on Remote Engineering and Virtual Instrumentation (REV), with 28 publications, in total (period 2004–2021).
- The number of citations per year, in Google Scholar and Web of Science (WoS). The graph contains several peaks, meaning one cannot identify one single publication as the seminal publication about VISIR. However, Gustavsson *et al.* (2009) may be said to stand out, as it created a peak in both Google Scholar and Web of Science, in 2009, and it is the publication with more references in Google Scholar, whereas in 2011, not one but several publications contributed to the highest number of citations in Google Scholar (approx. 450 citations).
- The citations vs. sources in WoS and Google Scholar, respectively. In the first graph (Google Scholar), REV stands out as the primary source, whereas in the second graph (WoS), IEEE Transactions on Learning Technologies (TLT) appears as the source with the highest number of citations, as already mentioned in the previous bulleted item.

Finally, Silva *et al.* (2021) present the full author network (without outliers), which evidences the strong collaboration bonds among those who have published on the subject of VISIR. Figure 4.9 is a complete version of that full author network, as it also includes the outliers, i.e., those authors with one publication on VISIR.

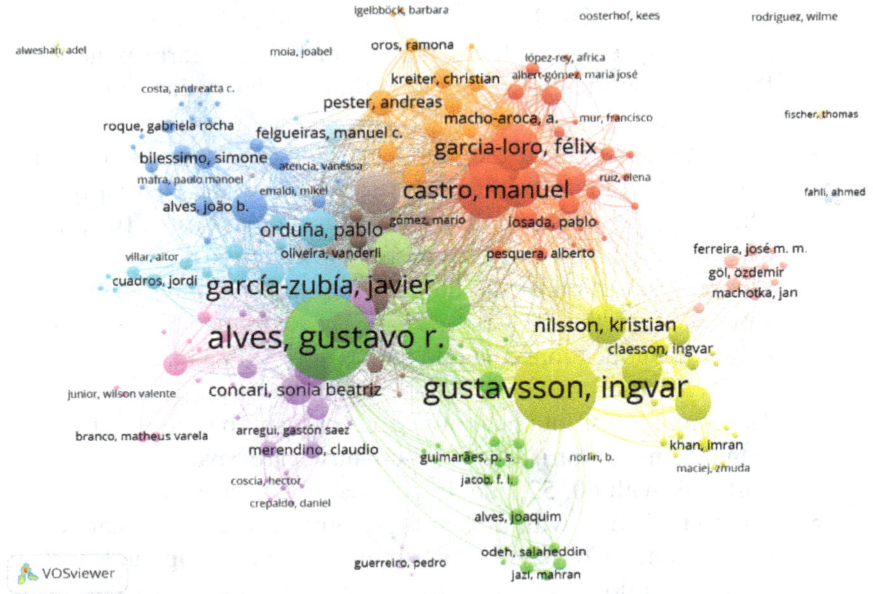

Figure 4.9. The full author list taken from the VISIR bibliographical database.

4.5 Conclusion

This chapter sought to introduce Ingvar Gustavsson, the creator of the VISIR remote laboratory, the history behind its development and distribution, and the community that grew around it. Summarizing, the chapter condenses a journey that started in 1999 and has spread to five continents, having served, to date, dozens of researchers, hundreds of teachers, thousands of students, and witnessed millions of remote experiments with electrical and electronic circuits.

References

Alves, G. R., *et al.* (2016). Spreading remote labs usage: A system — A community — A federation. In *Proceedings of the 2nd International Conference of the Portuguese Society for Engineering Education (CISPEE)*, Vila Real, Portugal.

Arguedas-Matarrita, C. (2017). Diseño y desarrollo de un Laboratorio Remoto para la enseñanza de la física en la UNED de Costa Rica. Doctoral Thesis. Universidad Nacional del Litoral.

Fischer, T. (2007). An overview of the VISIR open source distribution 2007 and ideas for further developments. In *Proceedings of the 4th International Conference on Remote Engineering and Virtual Instrumentation (REV)* (Vol. 3, pp. 1–2), Porto, Portugal.

García-Loro, F. (2018). Evaluación y Aprendizaje en Laboratorios Remotos: Propuesta de un Sistema Automático de Evaluación Formativa Aplicado al Laboratorio Remoto VISIR. Doctoral Thesis. Universidad Nacional de Educación a Distancia.

GOLC (2021). Retrieved from http://online-engineering.org/GOLC_online-lab-award.php.

Gustavsson, I. (2001). Laboratory experiments in distance learning. In *International Conference on Engineering Education* (pp. 14–18), Oslo, Norway. Retrieved from http://openlabs.bth.se/static/igu/Publ/Konferensbidrag/PaperICEE01.pdf.

Gustavsson, I. (2002a). Remote laboratory experiments in electrical engineering education. In *ICCDCS 2002 — 4th IEEE International Caracas Conference on Devices, Circuits and Systems*, Aruba, Venezuela. https://doi.org/10.1109/ICCDCS.2002.1004082.

Gustavsson, I. (2002b). A remote laboratory for electrical experiments. In *Proceedings of the 2002 American Society for Engineering Education Annual Conference & Exposition* (pp. 11285–11293). https://doi.org/10.18260/1-2--10624.

Gustavsson, I. (2003a). User defined electrical experiments in a remote laboratory. In *Proceedings of the 2003 American Society for Engineering Education Annual Conference & Exposition* (pp. 8.1233.1–8.1233.10), Nashville, US, ASEE Conferences. https://peer.asee.org/12628.

Gustavsson, I. (2003b). Remote laboratory for electrical experiments. In T. A. Fjeldly & M. S. Shur (Eds.), *Lab on the Web: Running Real Electronics Experiments via the Internet* (pp. 175–219), John Wiley & Sons, Inc., Hoboken, New Jersey. https://onlinelibrary.wiley.com/doi/epdf/10.1002/0471727709.fmatter

Gustavsson, I. (2003c). A traditional electronics laboratory with Internet access. In *Proceedings of the International Conference on Networked e-Learning for European University of Granada, Spain.*

Gustavsson, I. (2003d). A remote access laboratory for electrical circuit experiments. *International Journal of Engineering Education*, 19(3), 409–419.

Gustavsson, I. (2003e). Traditional laboratory exercises and remote experiments in electrical engineering education. In *International Conference on Engineering Education* (pp. 1–7), Valencia, Spain. Retrieved from http://www.its.bth.se/staff/igu/.

Gustavsson, I., Zackrisson, J., Åkesson, H., Håkansson, L., Claesson, I., & Lagö, T. (2006). Remote operation and control of traditional laboratory equipment. *International Journal of Online Engineering*, 2(1), 1–8. Retrieved from https://online-journals.org/index.php/i-joe/article/view/326/0.

Gustavsson, I., *et al.* (2009). On objectives of instructional laboratories, individual assessment, and use of collaborative remote laboratories. *IEEE Transactions on Learning Technologies*, 2(4), 263–274.

Håstad, M., Lundkvist, L., & Wikström, H. E. (2001). DAL and MAL — two projects to improve distance learning in the technical field in Sweden. *International Journal of Innovation in Science and Mathematics Education*, 7(1), 1–7.

Hernández-Jayo, U. (2012). Metodología de Control Independiente de Instrumentos y Experimentos para su Despliegue en Laboratorios Remotos. Doctoral Thesis. Universidad de Deusto. Retrieved from https://dkh.deusto.es/comunidad/thesis/recurso/metodologia-de-control-independiente-de/d9872dca-b743-4e3b-9791-2aae0cc31bbf.

IAOE (2021). International Association of Online Engineering — VISIR Federation. Retrieved from http://online-engineering.org/VISIR-Federation_about.php.

ICo-op (2012). Retrieved from https://web.archive.org/web/20160306012553/; http://ico-op.eu/.

Internet Archive (2021). About the Internet Archive. Retrieved from https://archive.org/about/.

Najimaldeen, R. (2017). A Federation of Online Labs for Assisting Science and Engineering Education in the MENA region. Doctoral Thesis. Universidade do Algarve.

PILAR (2016). Retrieved from http://www.ieec.uned.es/pilar-project.

Rosén, A. & Nilsson, K. (2007). Control system for a remote electronics laboratory. Blekinge Institute of Technology. Retrieved from http://urn.kb.se/resolve?urn=urn:nbn:se:bth-3296.

Silva, M., Fidalgo, A., Marques, M., Alves, G., Salah, R., & Jacob, F. (2021). A comprehensive VISIR bibliographical reference, *2021 World Engineering Education Forum/Global Engineering Deans Council (WEEF/GEDC)*, Madrid, Spain, 2021, pp. 468–475. doi: 10.1109/WEEF/GEDC53299.2021.9657332.

The SEFI Francesco Maffioli Award of Excellence for Developing Learning and Teaching in Engineering Education (2018). Retrieved from https://www.sefi.be/2018/10/02/prof-ingvar-gustavsson-receives-sefi-francesco-maffioli-award-for-teaching-excellence/.

VISIR+ (2015). Retrieved from https://www2.isep.ipp.pt/visir/.

Chapter 5

Pedagogical and Research Impacts of VISIR

5.1 Introduction

This chapter presents and discusses the research conducted to date on the pedagogical impact of VISIR. It starts by framing the research questions (or lines) underpinning the issue and then discusses, as exhaustively as possible, the various publications (i.e., articles in journals and conference proceedings, plus theses) that report on supporting tools, evidence, and findings stemming from this research.

After the coining of the expression "Second-Best-to-Being-There" (SBBT) by Aktan *et al.* (1996), in reference to a remote laboratory developed for experiments in control, several researchers started to investigate the educational value of remote experimentation. Although this very expression — "Second-Best" — implies a diminished value of remote laboratories in relation to traditional hands-on laboratories (TL), some authors, for instance, Brinson (2015), have argued that non-traditional laboratories (NTL) (i.e., remote and virtual laboratories) support student learning outcome achievement as well as TL. Ma and Nickerson (2006) also observe that the boundaries between the three laboratory types (hands-on, virtual, and remote) are blurred in the sense that most laboratories are mediated by computers and that the psychology of presence may be as important as technology. In fact, many laboratory experiments are accessed through instruments controlled by a computer, creating what Soysal (2000) describes as "computer-integrated experimentation (CIE)", which can be delivered on-site or online. In Soysal's opinion, on-site CIE can be considered the most effective experimental activity, which is in line with the opinion of Aktan *et al.* (1996).

Thus, the question is as follows:

(RQ1) *Is VISIR equally effective, in terms of achieving a given learning outcome associated with experiments with electrical and electronic circuits, as doing the same experiments in a hands-on lab or in a virtual lab (i.e., circuit simulator)?*

Considering that VISIR is a real (remote) laboratory mediated by a computer interface and that every student interaction with the laboratory is recordable and traceable, the previous question(s) also triggers two additional questions. These questions are as follows:

(RQ2) *Is it possible to build a learning analytic tool that helps instructors to realize whether VISIR is helping students to understand how to conduct experiments with electrical and electronic circuits?*, and

(RQ3) *If — with VISIR — students are using a computer-mediated interface remotely to perform experiments with real instruments and components, how do they distinguish it from running simulations, in a virtual laboratory, which use computer models, instead of real artifacts? In other words, how do students distinguish VISIR from a virtual laboratory?*

The following three sections provide an overview of the research conducted while addressing each of these questions, respectively.

5.2 Measuring the Educational Impact

To provide an answer to the first question (RQ1), we will search the VISIR bibliographical database for all publications that focus on (i.e., a perfect match) or address (i.e., an acceptable contribution) the educational impact of VISIR.

Following a chronological perspective, the first publication that addressed the educational impact of VISIR dates from 2009 (Gustavsson *et al.*, 2009), in the form of a students' satisfaction questionnaire with 17 closed questions. Among the findings are the following: "*VISIR is accepted as a good learning tool*"; "*it is very important to improve the sense of reality in the students when they use any remote lab. This is especially important in VISIR because it is real, but it seems to be a simulator*"; and "*students think that VISIR is more useful than usable, that is, the students cannot exploit all the potential of VISIR because its usability is not high*".

The following year, Claesson *et al.* (2010) presented an article on the educational impact of using VISIR with students from secondary school level. The work was carried out within the scope of Lena Claesson's MSc dissertation and involved comparing the use of the VISIR remote laboratory and the use of a traditional hands-on laboratory. Again, the remote laboratory work was evaluated with a questionnaire, this time with eleven closed questions and two open questions. Among the findings, Claesson *et al.* (2010) reported that

> The majority of the students were satisfied. The students showed great interest in the laboratory experiments and appreciated that it was not simulations but happened in real life. Although a few students did not realize that it was real experimental work and they wondered why they got different result from measurements, in a sequence, when they use the Perform Experiment button, ...

In 2011, the number of publications focusing on and addressing the educational impact of VISIR increased to four. García-Zubía *et al.* (2011) employed the same students' satisfaction questionnaire over three consecutive academic years (2008–09, 2009–10, and 2010–11), with three sets of questions focusing on the usefulness, sense of reality/immersion, and usability of VISIR. This study included students using both VISIR and a traditional hands-on laboratory, and students using only VISIR. Based on the questionnaire results, authors concluded the following:

> (1) VISIR [...] is a functional and useful learning instrument; (2) teacher experience at VISIR plays the crucial role in its integration into student experimentation activities; (3) [...] the sense of immersion of the remote experiments should be improved; (4) best knowledge and skills [...] students obtain from combination of experiments at traditional laboratory and remote lab (VISIR).

Alves *et al.* (2011) and Costa-Lobo *et al.* (2011) presented, for the first time, evidence of the educational impact of VISIR, supported by a suitably designed assessment methodology and a large cohort of more than 500 students. The students were divided into two natural groups: one group attending both hands-on and remote laboratory classes and another group simply using VISIR. Analysis of the pre- and post-questionnaires, correlated with the number of accesses to VISIR, provided the following conclusions: First, that "*Concerning the students learning and development of competences, the majority of students presented gains between 0 and 40%, and students who simply used VISIR presented similar or better results than those who used VISIR and had hands-on classes*", and second that "*the initial belief that VISIR would help students improve their ability to set up electrical circuits, in hands-on classes, was clearly confirmed by the students' perceptions reflected in the final questionnaire and the teachers' feedback provided during their interviews*". Finally, that same year, Viegas *et al.* (2011) expanded the work reported in the two aforementioned publications and presented results based on the use of VISIR in six different courses, of different sizes (47 to 574 students enrolled), from six different degrees, therefore, with quite different backgrounds. In total, more than 1200 students participated in the study. Among the findings, authors mentioned the following:

> In scientific courses (S1, S2, and S3) the head teachers have no doubt that students have developed some laboratory competences while using VISIR, namely their ability to judge the validity of their results. However, they also state that it would be more helpful in Introductory Courses, while working on principles of measuring, interconnecting components and measurement equipment. In the four courses where VISIR has been more systematically used (S1, S2, S3, and C1), providing a longer accompaniment with the system, the analyzed gains in the competence test (competences worked with VISIR analysis) were greater than in those where this task was unique (C2 and B1). Even so, the gain was not very high, as can be seen […] in C1 course.

Some of the works published in 2011 were continued in the following year and reported in Claesson and Håkansson (2012), Fidalgo *et al.* (2012), and Alves *et al.* (2012). In 2013, Claesson *et al.* (2013) extended the work reported in Claesson and Håkansson (2012), with data from the 2012–2013 academic year. In 2014, García-Zubía *et al.* (2014) reused the pre- and post-questionnaires presented in Alves *et al.* (2011), with 87 students, and obtained a total of 188 completed questionnaires (including some filled-in by teachers). This work contains a brief analysis of questionnaire reliability, where authors present evidence (Cronbach's alpha) supporting the need to improve some of the questions included in the questionnaire. This need is addressed in a later work described in Cuadros

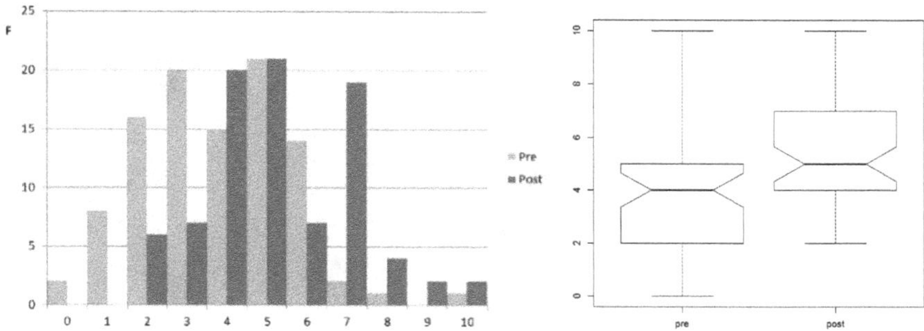

Figure 5.1. Pre- and post-test questionnaire results reported in García-Zubía *et al.* (2014).

et al. (2021). Nevertheless, the results obtained, illustrated in Figure 5.1, show the positive impact of VISIR in the students' learning process. These results were submitted to the Wilcoxon's signed-rank test and the student's *t*-test, where both return a $p < 0.001$, which sustains the conclusion that the result of the post-questionnaire is higher than the result of the pre-questionnaire. Also in 2014, Marques *et al.* (2014) published a larger study based on the same target courses of Viegas *et al.* (2011). The study addressed a somewhat different research question, i.e., "*Is VISIR always useful, no matter how it is integrated into a course? Or are there certain conditions/characteristics that enhance student learning?*", yet the term "useful" is understood as "*the degree to which VISIR-related learning outcomes were accomplished by the students*", which is related to the educational impact of VISIR. Authors concluded that "*VISIR is shown always to be of benefit for more motivated students, but this benefit can be maximized under particular conditions and characteristics*", while also presenting those conditions and characteristics. Again, by using pre- and post-questionnaires, the educational impact of VISIR is visible in Figure 5.2, which illustrates the learning gain results measured through the students' answers to questions related to laboratory competences.

Although Figure 5.2 relates to one specific course (*S*2), the advantages of using VISIR become evident by comparing the learning gains of students who used it (circle) against those who did not use it (X). Also in 2014, two more publications reported on the subject. Viegas *et al.* (2014) described the usage of VISIR, in parallel with simulations, and hands-on, to improve students' experimental competences; and Odeh *et al.* (2014) presented additional evidence on the usability of VISIR in a case study involving 71 students from the Al-Quds University, in Jerusalem, who used a VISIR system installed at the Polytechnic of

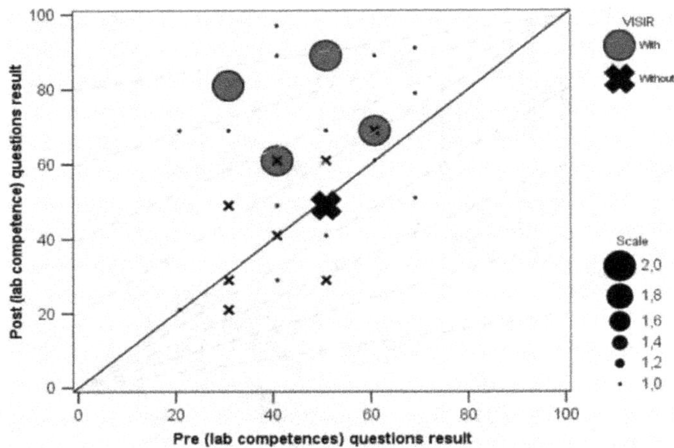

Figure 5.2. Learning gain results in laboratory competence-related questions, concerning course *S2*, as reported in Marques *et al.* (2014).

Porto — School of Engineering, in a clear example of international collaboration. Using a student survey questionnaire, the authors were able to measure a positive response concerning the usability of VISIR, with an average result of 4.00 on a 5-point Likert scale, which contrasted with an average result of 2.20 points, concerning the students' satisfaction with the hands-on laboratory.

In 2015, Romero *et al.* (2015) initiated a new research line towards the development of an automatic assessment model, able to (1) scaffold students' learning with VISIR, by reporting common errors and mistakes, and (2) inform teachers about students' learning progress. The proposed model used data automatically collected from students' interaction with VISIR, a list of expected (correct) results for every experiment provided by the teacher(s) and learning analytics techniques. This work later evolved into the VISIR dashboard and forms a clearly defined research direction, as described in the following section.

In 2016, Lima *et al.* (2016a) expanded the work reported in Viegas *et al.* (2014). The latter reported on the use of VISIR for experimenting with electric circuits fed with Direct Current (DC), while the former reported on experiments with electric circuits fed with Alternate Current (AC). The paper presented the Learning Objectives (LO) associated with this new syllabus item and covered the following research question: "*Is the simultaneous usage of different online lab resources — the VISIR remote lab and a virtual lab — useful to support students learning in general and/or for developing their experimental competences?*". The paper then reported evidence — table II — of a significant correlation between the number of accesses to several resources and the grades obtained, namely in C2 calculus and total test outcome. The work with the virtual laboratory (online simulator) seemed to help students develop important competences in terms of calculus, while the VISIR remote laboratory helped students in a broader manner, helping them to achieve better performances in general. These results suggest that some of these online resources did help students to learn, even though they did not guarantee their performance in each component.

Also in 2016, Alves *et al.* (2016) reused the statistical methods employed in Lima *et al.* (2016a) further to expand the analysis of the data reported in Viegas *et al.* (2014) and presented new evidence corroborating the findings that

> Students who frequently used virtual resources show a higher level of performance *[and]* class attendance seems to have particular importance in the result obtained in the simulation/remote lab components, perhaps because it is crucial that they overcome the first difficulties in the presence of the teacher, in class.

The first consortium meeting of the VISIR+ project, held at BTH in February 2016, assembled a large group of researchers who later focused their work on analyzing the educational impact of VISIR in the five institutions of Latin America (Argentina and Brazil) that received this remote laboratory. One of the first tasks was to compile all previous works that contained information about the educational impact of VISIR. The results were published in Lima *et al.* (2016b, 2016c). As stated by the authors:

> The purpose of this work is to understand and systematize the scientific research using VISIR's approach, done so far. The tackled research question is: *Considering VISIR implementation and usage reported in literature until May 2016, which common outcomes and indicators of consistent results can be found* ***in the different didactical approaches****?*

Lima *et al.* (2016b) identified 54 papers, with the majority (52%) devoted to technical issues. By thoroughly examining the other part (48%), they identified a total of 22 courses, covering more than 4400 students, from different educational levels, i.e., Secondary, Vocational, and Higher. In response to their initial research question, authors concluded the following:

> **VISIR is a functional and useful learning instrument**, well accepted by students, which should be used as a complement to hands-on lab or as a tool for distance learning. It improves students' competence and knowledge as it is reported in 59% of the analyzed cases.

Also in 2016, García-Zubía *et al.* (2016) expanded the work reported in García-Zubía *et al.* (2014), using additional data from two academic years (2013–2014 and 2014–2015) to address the following research question: "*Does the use of VISIR* [...] *have a positive effect on students' learning process?*" The authors concluded the following:

> The main conclusion, based on the use of an O-X-O design study, is that **using VISIR** in basic electronics education **helps students in their learning** and has a positive effect. This conclusion is statistically significant and was valid for the five different student groups on two different courses, in three different cities, with three different teachers and two different educational levels.

More evidence regarding the educational impact of VISIR was published in 2017. Many of those works were carried out within the scope of the VISIR+ project, involving students and teachers from Argentina and Brazil, and a data collection approach plus a set of pedagogical evaluation tools commonly defined by the project partners. Some of the pedagogical tools are described in Pozzo *et al.* (2017) and, later, in Pozzo (2019), while Viegas *et al.* (2017) present the preliminary results of the teachers' training actions and the first implementations in the five Latin American (LA) partner institutions (see Table 5.1). Finally, Lima *et al.* (2017a) present Lima's PhD thesis project, which used the didactical implementations employed in the VISIR+ project as the primary data source for her study.

The VISIR+ project ended in April 2018, generating a large quantity of data from the didactical implementations employed in the LA partners. While most of the analyses are compiled in Lima's PhD thesis, some implementations form the subject of a few articles published in 2018 and 2019, namely Viegas *et al.* (2018) and Pavani *et al.* (2018) — data from implementations at PUC-Rio; Lima (2018) — data from implementations at one associated partner of PUC-Rio, i.e., the Catholic University of Petrópolis, RJ, Brazil; and Marchisio *et al.* (2018) — data from implementations at UNR, among others.

Table 5.1. Didactical implementations in each LA partner institution (Reprinted from Viegas *et al.*, 2017).

Courses	LA Partner	Teachers Team	Students	VISIR's Usage
Calculus IV	UFSC	3	40	Sporadic
Probability and Statistics		1	50	Sporadic
Electronics II	IFSC	1*	13	Frequent
Basic Electronics			13	Frequent
Amplifying Structures			10	Frequent
Electric Circuits I		1	31	Frequent
Electric Circuits I		1	40	Frequent
Electricity I		1	50	Frequent
Electric and Electronic Circuits	PUC-Rio	1	18	Frequent
Complementary Activity		1	Eng. students	Continuous
Physics of Devices	UNR	2	17	Sporadic
Electronics I (+)	UNSE	4*	15–20	Frequent
Electronics II (+)			15–20	Frequent

Notes: *Same teachers team. (+) To be implemented soon.

Finally, in 2020, Natércia Lima defended her PhD thesis, presenting the analyses of how VISIR was used in 26 different courses, involving 52 different teachers and 1794 students. In her work, Lima lists the many aspects that should be considered when planning to use VISIR as an effective educational tool, aiming at improving students' experimental competences. One year later, in 2021, a partial view of the research conducted was reported in Lima *et al.* (2021).

5.3 A Dashboard for VISIR

As mentioned earlier, Romero's PhD thesis (Romero, 2015) contains the foundations that led to the development of the VISIR Dashboard (VISIR-DB). Quoting from the introduction: "*Is it possible to build a learning analytic tool that helps instructors to realize whether VISIR is helping students to understand how to conduct experiments with electrical and electronic circuits?*" The answer to this question is yes, and it corresponds (on an initial development state) to the Activities Automatic

Assessment System (AAAS) model proposed and described in Romero *et al.* (2015) and Romero (2015).

According to Romero (2015), the AAAS model uses (1) data extracted from VISIR; (2) learning analytics techniques; and (3) a set of evaluation rubrics, i.e., information provided by the teachers on how a given experiment should be correctly performed and how any deviations should be categorized. The data extracted from VISIR correspond to the circuit set-up and the parameters established in each of the connected test and measurement devices, i.e., the DC power supply, the function generator, the multimeter, and the oscilloscope. The learning analytic techniques are used at two different levels: one corresponding to circuit-level information, i.e., (i) how well students executed a given experiment, and another corresponding to system-level information, i.e., (ii) how students interacted with VISIR.

The first information level (i) is described in more detail in Romero (2015) and, later, in García-Loro (2018), corresponding to García-Loro's PhD dissertation, which may be considered an expansion of Romero's PhD work. In García-Loro (2018), the concept of a "model pattern" is introduced, referring to the correct mounting of the circuit under experimentation plus the connections with the test and measurement instruments and their correct configuration. For each circuit/experiment, depending on the given learning goal, a model pattern needs to be defined, using input from the teacher(s), and stored in the evaluation system, to classify a given experiment as correct/incorrect — this corresponds to input (3) of the AAAS model specified by Romero (2015).

The second information level (ii) is missing from Romero's PhD work, which may be seen as a proof of concept, with many functionalities implemented in a manual (offline) way, rather than in an automatic (online) way, as intended. In fact, the whole concept is further developed by a multidisciplinary team (including experts in statistics and pedagogy) and presented in García-Zubía *et al.* (2019a), which describes the VISIR-DB, an app designed and developed with Shiny (Chang *et al.*, 2018). The Shiny app was originally arranged in four sections, each one organized into several subsections and panels, in which meaningful visualizations are displayed:

(1) The Data Input section, which makes it possible to select a .CSV file — corresponding to the recording of the interactions in the WebLab-Deusto platform — to be analyzed. Summary data as the number of users or date range are provided for double-checking purposes (see Figure 5.3).

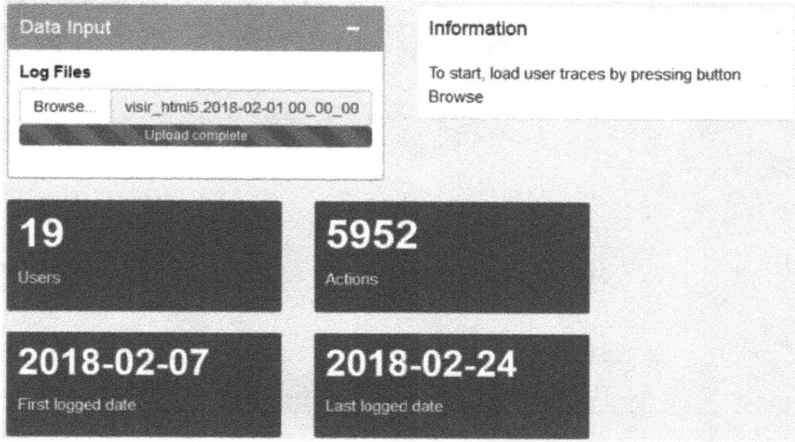

Figure 5.3. VISIR-DB — Data input section. Reprinted from García-Zubía *et al.* (2019a).

(2) The Global Results section, which provides different summaries and visualizations intended to provide a summarized view of a group of students, both in terms of time and amount of work, i.e., number of actions and experiments performed. For instance:

- Figure 5.4 corresponds to a histogram of time on VISIR, per student. In this example, the total time was 80.81 h, with an average time, per student, of 4.25 h (19 students).
- Figure 5.5 provides an overview of the temporary distribution of the students' work. Rows correspond to days, columns to individual students, and the color intensity corresponds to time on task.
- Figure 5.6 corresponds to a histogram of number of circuits, per student. In this example, a total of 5077 circuits were set up, with an average number of 267 circuits per student.
- Figure 5.7 is the equivalent of Figure 5.6, after normalizing the various circuits set by the students, i.e., the VISIR-DB can detect two equivalent circuits, with the same components and connections (including the test and measurement instruments), and convert them into one normalized circuit.
- Figure 5.8 shows students' work (number of normalized circuits) compared with time on task (time spent using VISIR). This chart makes it easy to spot students who may have difficulties with the task or the class.

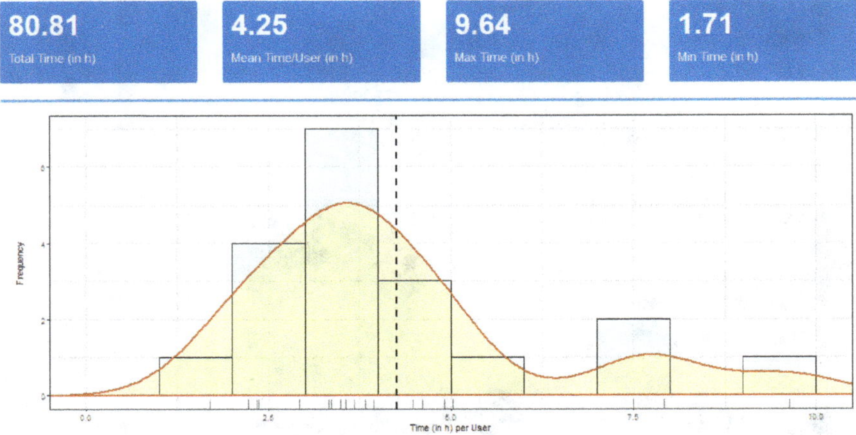

Figure 5.4. VISIR-DB — Global results section — histogram of time (h) per student. Reprinted from García-Zubía *et al.* (2019a).

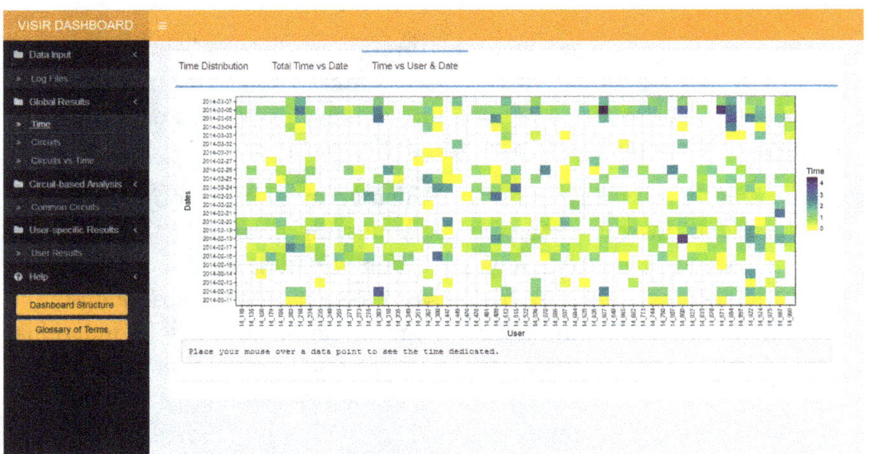

Figure 5.5. VISIR-DB — Global results section >> Time: Temporary distribution of the students' work. Rows are dates and columns are users, the color scale indicates time on task. Reprinted from García-Zubía *et al.* (2019a).

(3) The Circuits-Based Analysis section, which includes the list of circuits implemented by the users and shows who built each of the circuits (and when). In this section, one can select a normalized circuit and see how many times it was set up by each student. In the example illustrated in Figure 5.9, the circuit corresponding to

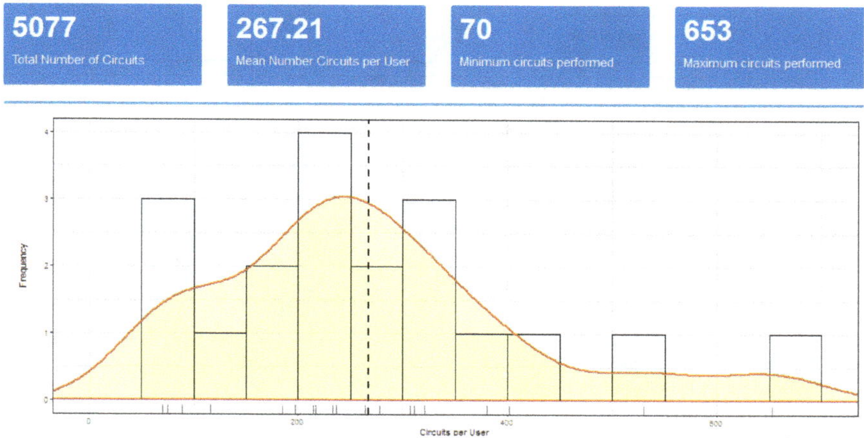

Figure 5.6. VISIR-DB — Global results section — Histogram of the number of circuits per user. Reprinted from García-Zubía *et al.* (2019a).

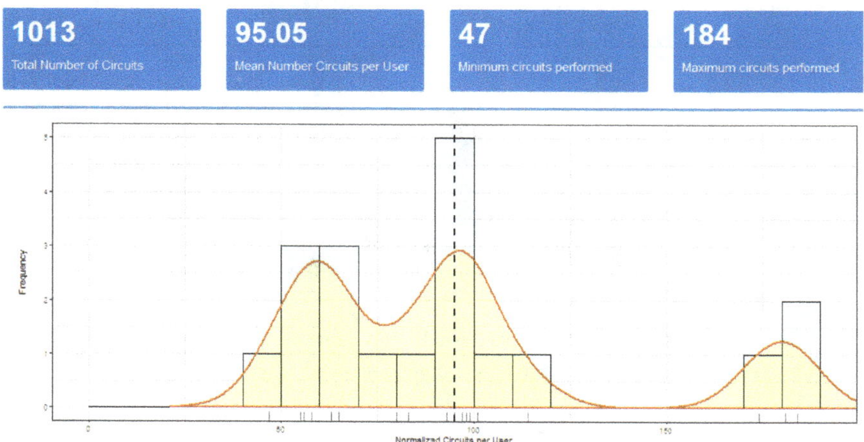

Figure 5.7. VISIR-DB — Global results section — histogram of the number of (normalized) circuits per user. Reprinted from García-Zubía *et al.* (2019a).

measuring the resistance of a 1-kΩ resistor was not built by students 1 and 17, while student 9 implemented it 18 times (columns correspond to students).

(4) The User-Specific Results section, which provides information about a specific user and the circuits they created. For instance, Figure 5.10 provides a graphical overview of every type of circuit/experiment

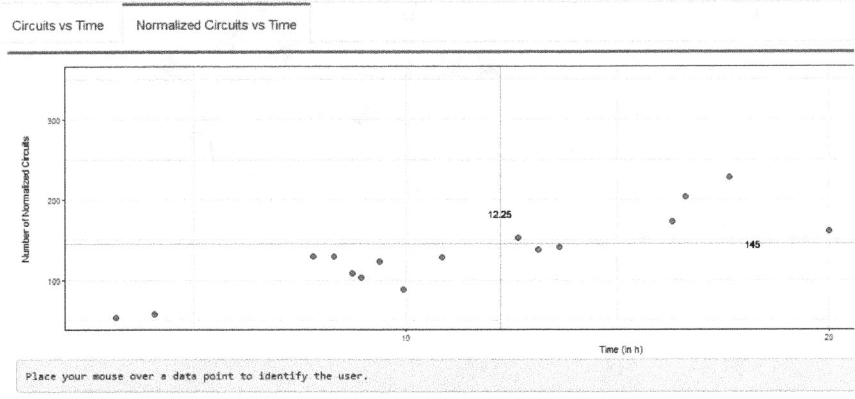

Figure 5.8. VISIR-DB — Chart with the number of normalized circuits vs. time. Reprinted from García-Zubía *et al.* (2019a).

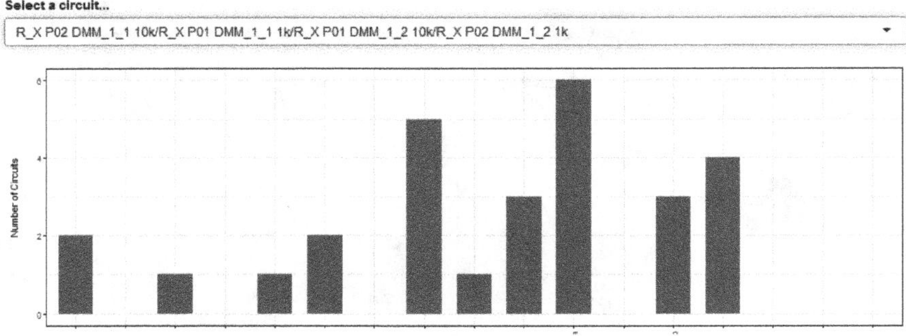

Figure 5.9. VISIR-DB — Circuit-based Analysis section — creation of one (normalized) circuit by each student. Reprinted from García-Zubía *et al.* (2019a).

(measure current — red, measure resistance — green, measure voltage — blue, and circuit/configuration with errors — yellow) performed by each of the 19 students.

While the initial VISIR-DB version allowed teachers better to understand how students were performing with VISIR, it did not provide feedback to the students on specific errors made nor did it help teachers to understand the sort of errors students were mostly making. Adding this functionality to the VISIR-DB requires integration of García-Loro's PhD work, in particular the generation of a model pattern for each circuit/experiment. These models can be generated in a manual or automatic way.

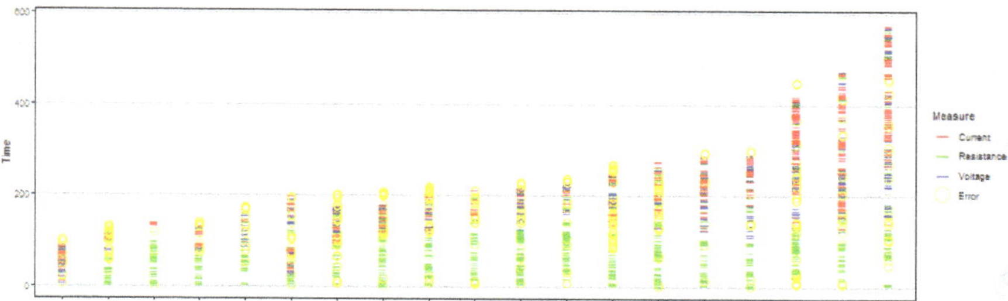

Figure 5.10. VISIR-DB — Circuit-based Analysis section — creation of one (normalized) circuit by each student. Reprinted from García-Zubía *et al.* (2019a).

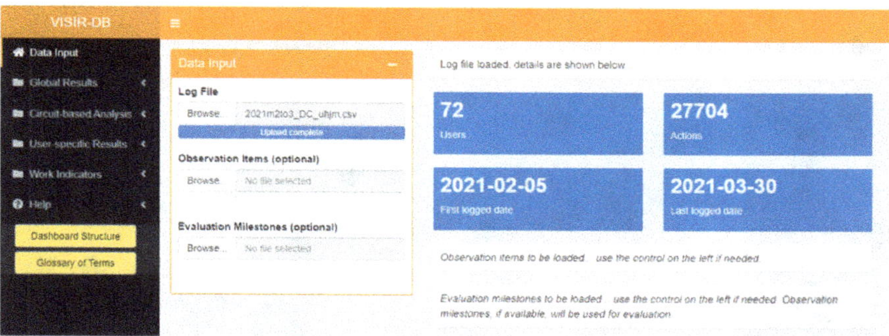

Figure 5.11. Front page of the new VISIR-DB version. Reprinted from Serrano *et al.* (2022).

The first option, i.e., the manual approach, is followed in a more recent work, reported in Serrano *et al.* (2022), which describes the newest version of the VISIR-DB. The new VISIR-DB version includes a new section, named Work Indicators, which permits automatic assessment of each student's action by checking the latter against a set of observation items defined by teachers. Additionally, the user's work can also be automatically assessed using evaluation milestones, which are logical computations from the performances of each user to the observation items. These two new functionalities require two input files (i.e., an observation items file and an evaluation items file) that can be uploaded on the front page of the new VISIR-DB version, illustrated in Figure 5.11.

Work indicators can now be visualized in the corresponding new section of the VISIR-DB both as the group performance, percentage of students reaching each indicator (Figure 5.12), and as individual performances, on a map showing which indicators are attained by each student (Figure 5.13).

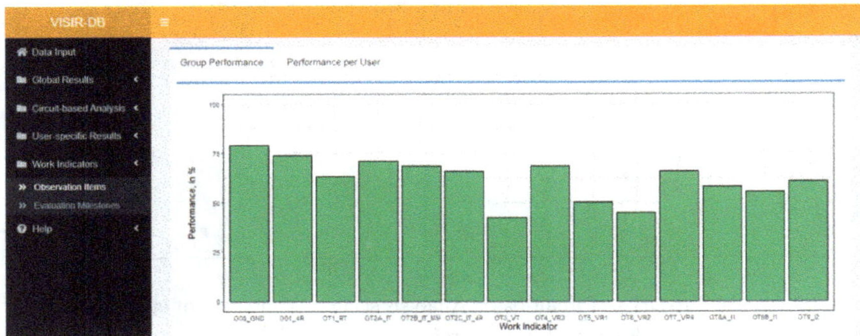

Figure 5.12. VISIR-DB — Work Indicators section — VISIR-BD group performance per observation item. Reprinted from Serrano *et al.* (2022).

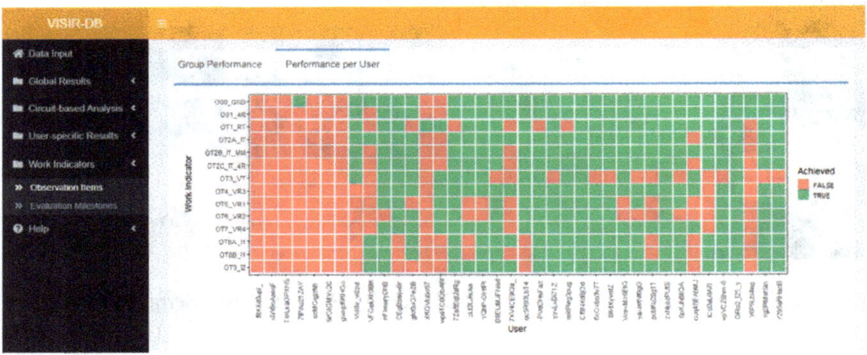

Figure 5.13. VISIR-DB — Work Indicators section — VISIR-BD individual performance per observation items and student. Reprinted from Serrano *et al.* (2022).

Another enhancement of the new VISIR-DB version is the support of simplified circuits, which were introduced in Cuadros *et al.* (2021). Briefly, a simplified circuit is a canonicalization of the built circuit that only considers what is connected to the multimeter and thus grasps the intended measurement of the student.

Finally, while García-Loro (2018) uses a method based on extracting information from experiments performed by the teacher to build a model pattern, he also points out some limitations to this method, i.e.,

> The greatest challenge [...] is to identify the experiment that the student is trying to design. Ergo, it is possible to identify the practical exercise

the student is experimenting with, but, as each one is composed of a set of experiments — an indeterminate but finite number of experiments — and each linked to a model pattern, **initially it is not known which model pattern should be used to check against the student's experiment**. This initial indeterminacy between the student's experiment and the model pattern entails a less specific and accurate feedback, specifically, generic feedback. Making it possible to identify at each experiment designed by the user which experiment they are trying to complete — and, consequently, which pattern model has to be used to check against — would allow a more exhaustive analysis, and a more specific feedback could be provided.

To address this limitation, several works aiming at defining specific and automatic feedback according to all possible errors linked to every possible experiment (considering the syllabus of courses that include experiments with electrical and electronic circuits and the learning goals associated with those experiments, in a course-specific context) have been published. The first work in this series corresponds to García-Zubía *et al.* (2019b) and proposes a generic framework for interpreting experimental errors in VISIR. While the proposed framework claims to be generic, its description is restricted to 1:1 circuits in DC mode, i.e., circuits with a DC power supply and a single resistor, plus a multimeter for measuring current–voltage (or resistance, assuming the resistor is not connected to the power supply). The second work corresponds to Mendonça *et al.* (2020) and expands the previous work to simple 1:1 circuits in AC mode, with either resistors or capacitors. The third work, by Sasdelli *et al.* (2021), updates the framework to very simple circuits including the most basic electronic component (with two leads), i.e., the diode. Finally, a more recent work described in Costa *et al.* (2022) addresses the use of intelligent classification algorithms automatically to characterize simple 1:1:0:1 and 1:0 circuits.

5.4 Differentiating Remote (Real) from Simulated Experiments

In this section, we finally address the third question, i.e., (RQ3) "*If — with VISIR — students are using a computer-mediated interface remotely to perform experiments with real instruments and components, how do they distinguish it from running simulations, in a virtual lab, which use computer models, instead of real artifacts? In other words, how do students distinguish VISIR from a virtual lab?*"

An alternative reading of this question may result in splitting it into two complementary dimensions, i.e., (i) *how students* ***perceive*** *VISIR as being different from a simulated environment* and (ii) *how teachers* ***demonstrate*** *that VISIR is a (real) remote lab, where real experiments, with real components and instruments, are performed.*

Lima *et al.* (2017b) have evaluated student perception of VISIR and simulation results, using qualitative and quantitative data obtained during the VISIR+ project. The analysis results indicate that a considerable number of students fail to distinguish the differences (i.e., mathematical models are not 100% accurate and hence do not entirely reflect the response of real-world systems or phenomena), even though they know their definition. Furthermore, the authors explain that this misperception does not appear to change according to the context (country, course, academic year, or course content), students' final grades, teachers' didactical approach, or even implemented tasks. This aspect, i.e., the correlation between the link to the physical world and the effectiveness of virtual vs. (real) remote laboratories, was also discussed in detail in Ma and Nickerson (2006), who sustain that "*belief may be more important than technology*". As Lima *et al.* (2017) have also collected qualitative data, it is possible to illustrate some of the students' responses that specifically illustrate their misconception, even though some of these students obtained high grades or were considered — by the teacher(s) — to have developed high order experimental skills. A student response that can inspire teachers to demonstrate that real experiments are performed in VISIR is as follows:

> *I have heard [of] VISIR a development of a graphical interface, however it* ***doesn't use physical connections like wire. In VISIR we can truly make a real circuit. VISIR does not require the use of physical means, such as wires.***

From a technical perspective, in an electrical circuit, wires are one of the major causes of a noise effect called "crosstalk". One way to increase this effect is to have a strong inducing signal, e.g., a high-frequency square wave, with an electrical current several times higher than the electric current flowing in the neighboring (induced) signal. The circuit illustrated in Figure 5.14 is a possible example.

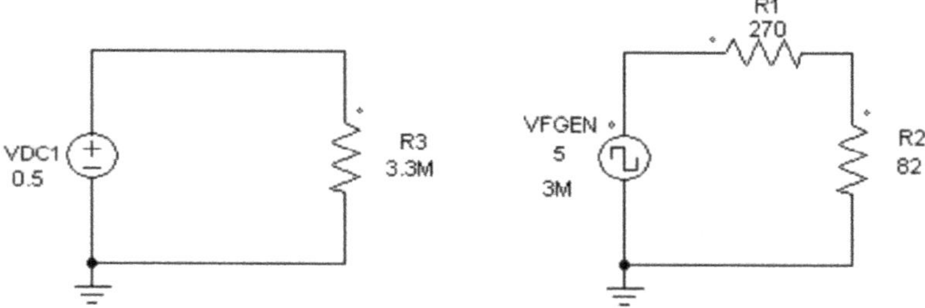

Figure 5.14. The proposed demo circuit(s). The induced circuit (left) and the inducing circuit (right).

The specifications of the left-side (induced) circuit are as follows:

- **DC source voltage**: 0.5 V
- **Load resistance**: 3.3 MΩ.

The specifications of the right-side (inducing) circuit are as follows:

- **AC source voltage, square wave**: 5 V_{pp}, 3 MHz, DC = 0 V
- **Load resistance**: 352 Ω.

Applying Ohm's law, the electric current on the left-side circuit is as follows:

$$I_{\text{left-side circuit}} = \frac{U}{R} = \frac{0.5}{3.3\times10^{6}} = 0.15\times10^{-6} = 0.15\,\mu\text{A}$$

And the electric current on the right-side circuit is as follows:

$$I_{\text{right-side circuit}} = \frac{U}{R} = \frac{5}{352} = 0.0142 = 14.2\times10^{-3} = 14.2\,\text{mA}$$

Therefore, the relation between the two electrical currents is as follows:

$$\frac{I_{\text{right-side circuit}}}{I_{\text{left-side circuit}}} = \frac{14.2\times10^{-3}}{0.15\times10^{-6}} = 94.67\times10^{3} \approx 1\times10^{5}$$

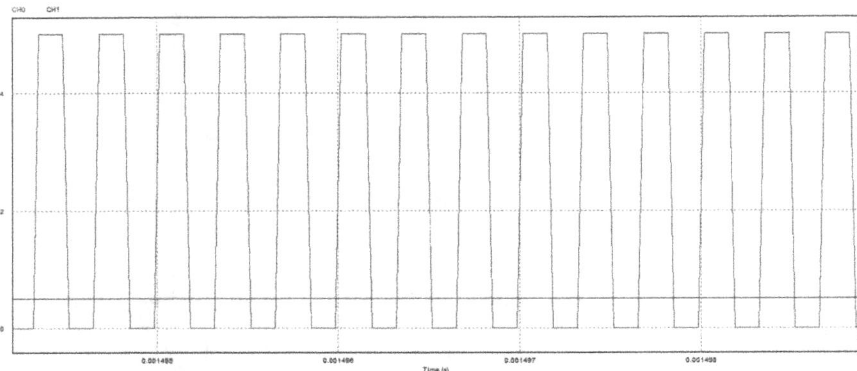

Figure 5.15. Simulation of the proposed demo circuit(s). The solid horizontal line represents the voltage drop at the 3.3 MΩ resistor.

This relation (1:100000) is large enough to make it possible to visualize the crosstalk effect. Note that while a simulation of the two circuits will not be affected by this crosstalk effect, the real (remote) experiment with real wires will be. Nonetheless, it would still be possible to simulate the crosstalk effect, but then the circuits would be different from the ones illustrated in Figure 5.14, as the wires would have to be modeled by a transmission line, i.e., a resistor in series with an inductance, with a small capacitance connecting to adjacent wires. To summarize, Figures 5.15 and 5.16 illustrate the differences between the simulated (modeled) and the remote (real) experiments, due to effects such as crosstalk, which can be used as an illustrative example for students struggling to understand those very same differences.

As an additional step, one could mention the fact that VISIR uses real instruments, for instance, a switched DC power supply. This sort of power supply presents ripple noise in the DC output because of the AC/DC conversion process. It is possible to visualize (and measure) the ripple noise, by using the oscilloscope, with the inducing circuit inactive, to suppress this second source of signal noise. Figure 5.17 illustrates the obtained waveforms obtained, with the oscilloscope input channel in DC coupling (left side) and in AC coupling (right side). The signal visualized in Figure 5.17 (left side) is clearly different from the signal visualized in Figure 5.16 (Channel 1 — blue line), even considering the vertical amplification to be different (Figure 5.16 — 1.0 V/div | Figure 5.17 — 0.5 V/div).

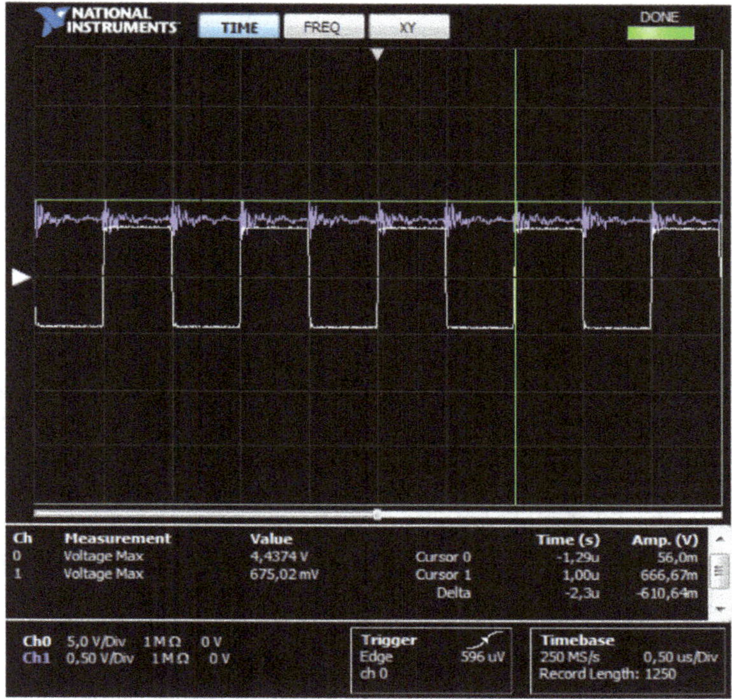

Figure 5.16. Visualizing the two signals in VISIR, with the oscilloscope. Blue line (with noise due to crosstalk) represents the voltage measured at the 3.3 MΩ resistor.

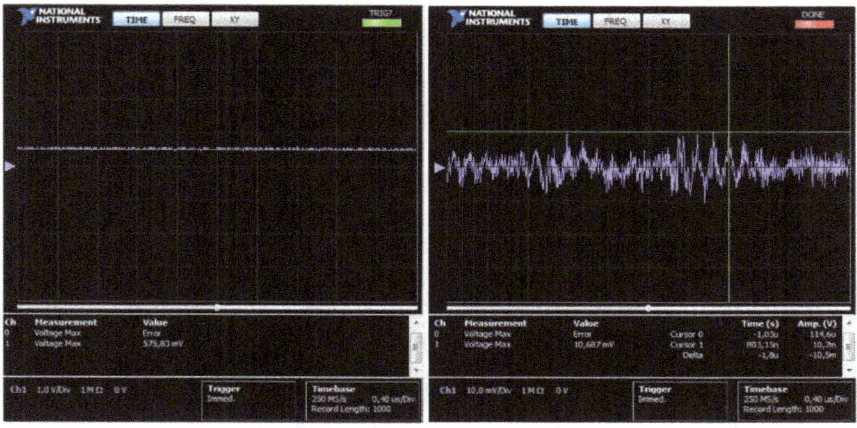

Figure 5.17. Visualizing the voltage drop at the 3.3 MΩ resistor, with the inducing circuit inactive. Oscilloscope — VISIR — in DC (left-side) and AC (right-side) coupling.

Summing up, how to differentiate simulations from real (remote) experiments done in VISIR has also been a research line explored by the VISIR community, as described in Branco *et al.* (2017a, 2017b).

5.5 Conclusion

The three RQs presented in Section 5.1 serve as a structuring guideline for this chapter, with each subsequent section (Sections 5.2–5.4) presenting published works that tackle each question (RQ1, RQ2, and RQ3). Those publications also present recommendations with regard to how better to use VISIR, by both teachers and students, covering many different educational scenarios (use inside/outside the classroom, use simultaneously with hands-on and virtual laboratories, etc.). The large number of scenarios addressed, and the intrinsic details associated with each scenario, makes it difficult to create a simple table summarizing all the recommendations produced, aiming at facilitating the teacher's task when designing an instructional plan incorporating the use of the VISIR remote laboratory. Furthermore, the evidence reported in the cited publications also allows teachers to build solid arguments, at an institutional level, vis-à-vis the advantages (and impact) associated with using VISIR in each course. This book (and this chapter in particular) is also meant to serve yet another piece of evidence maintaining the benefits of using VISIR to allow students to perform (real) remote experiments with electrical and electric circuits.

References

Aktan, B., Bohus, C. A., Crowl, L. A., & Shor, M. H. (1996). Distance learning applied to control engineering laboratories. *IEEE Transactions on Education*, 39(3), 320–326. https://doi.org/10.1109/13.538754.

Alves, G. R., Marques, A., Viegas, C., Costa-Lobo, M. C., Barral, R. G., Couto, R. J., … Gustavsson, I. (2011). Using VISIR in a large undergraduate course: Preliminary assessment results. In *Proceedings of the IEEE Global Engineering Education Conference (EDUCON), Amman, Jordan* (pp. 1125–1132). IEEE. https://doi.org/10.1109/EDUCON.2011.5773288.

Alves, G. R., Viegas, C., Lima, N., & Gustavsson, I. (2016). Simultaneous usage of methods for the development of experimental competences. *International Journal of Human Capital and Information Technology Professionals*, 7(1), 48–65. https://doi.org/10.4018/IJHCITP.2016010104.

Alves, G. R., Viegas, C., Marques, A., Costa-Lobo, M. C., Silva, A. A., Formanski, F., & Silva, J. B. (2012). Student performance analysis under different Moodle course designs. In *Proceedings of the 15th International Conference on Interactive Collaborative Learning (ICL)* (pp. 1–5), Villach, Austria. IEEE. https://doi.org/10.1109/ICL.2012.6402181.

Branco, M. V., Coelho, L. A., Schlichting, L. C. M., & Alves, G. R. (2017a). Differentiating simulations and remote (real) experiments. In *5th International Conference on Technological Ecosystems for Enhancing Multiculturality (TEEM)*, Cádiz, Spain. ACM Press. https://doi.org/10.1145/3144826.3145363.

Branco, M. V., Coelho, L. A., Alves, G. R., & Schlichting, L. C. M. (2017b). Aspectos de Diferenciação entre Laboratórios Remotos e Simuladores. In *XLV Congresso da Associação Brasileira de Ensino de Engenharia (COBENGE)*, Joinville, SC, Brazil.

Branco, M. V., Coelho, L. A., & Alves, G. R. (2017c). Estudo Comparativo entre Laboratórios Remotos e Simuladores. In A. L. Ferreira, A. Fidalgo & O. da Silva (Eds.), *TICAI (2017) TICs para el Aprendizaje de la Ingeniería* (pp. 117–123). IEEE Education Society.

Brinson, J. R. (2015). Learning outcome achievement in non-traditional (virtual and remote) versus traditional (hands-on) laboratories. *Computers in Education*, 87(C), 218–237. htpps://doi.org/10.1016/j.compedu.2015.07.003.

Chang, W., *et al.* (2018). Shiny: Web application framework for R. https://CRAN.R-project.org/package=shiny.

Claesson, L. & Håkansson, L. (2012). Using an online remote laboratory for electrical experiments in upper secondary education. *International Journal of Online and Biomedical Engineering (IJOE)*, 8(S2), 24. https://doi.org/10.3991/ijoe.v8is2.1941.

Claesson, L., Khan, I., Zackrisson, J., Nilsson, K., Gustavsson, I., & Håkansson, L. (2013). Using a VISIR laboratory to supplement teaching and learning processes in physics courses in a Swedish Upper Secondary School. In O. Dziabenko & J. García-Zubía (Eds.), *IT Innovative Practices in Secondary Schools: Remote Experiments* (pp. 141–176). Bilbao: Deusto University Press.

Costa, H. M., Alves, G. R., da Silva, J. B., & da Mota-Alves, J. B. (2022). Frequency detection of experimental errors through Learning Analytics techniques. In *Proceedings of the XV Congreso de Tecnología, Aprendizaje y Enseñanza de la Electrónica (XV Technologies Applied to Electronics Teaching Conference)* (pp. 1–6). https://doi.org/10.1109/TAEE54169.2022.9840595.

Costa-Lobo, M. C., Alves, G. R., Marques, A., Viegas, C., Barral, R. G., Couto, R. J., ... Gustavsson, I. (2011). Using remote experimentation in a large undergraduate course: Initial findings. In *Proceedings of the Frontiers in Education Conference (FIE)* (pp. 1–7). https://doi.org/10.1109/FIE.2011.6142913.

Cuadros, J., Serrano, V., García-Zubía, J., & Hernández-Jayo, U. (2021). Design and evaluation of a user experience questionnaire for remote labs. *IEEE Access*, 9, 50222–50230. https://doi.org/10.1109/ACCESS.2021.3069559.

Cuadros, J., Serrano, V., Lluch, F., García-Zubía, J., & Hernández-Jayo, U. (2021). Mapping VISIR circuits for computer-assisted assessment. In *2021 World Engineering Education Forum/Global Engineering Deans Council (WEEF/GEDC)* (pp. 524–527). IEEE. https://doi.org/10.1109/WEEF/GEDC53299.2021.9657349.

Fidalgo, A., Alves, G. R., Marques, A., Viegas, C., Costa-Lobo, M. C., Hernández-Jayo, U., … Gustavsson, I. (2012). Using remote labs to serve different teacher's needs — A case study with VISIR and remotelectlab. *International Journal of Online Engineering*, 8(Special Issue 3), 36–41. https://doi.org/10.3991/ijoe.v8iS3.2259.

García-Loro, F. (2018). Evaluación y Aprendizaje en Laboratorios Remotos: Propuesta de un Sistema Automático de Evaluación Formativa Aplicado al Laboratorio Remoto VISIR. Universidad Nacional de Educación a Distancia.

García-Zubía, J., Cuadros, J., Romero, S., Hernández-Jayo, U., Orduña, P., Guenaga, M., … Gustavsson, I. (2016). Empirical analysis of the use of the VISIR Remote Lab in teaching analog electronics. *IEEE Transactions on Education*, 60(2), 149–156. https://doi.org/10.1109/TE.2016.2608790.

García-Zubía, J., Gustavsson, I., Hernández-Jayo, U., Orduña, P., Angulo, I., Dziabenko, O., … López-de-Ipiña, D. (2011). Using VISIR experiments, subjects and students. *International Journal of Online Engineering*, 7(Special Issue 2), 11–14. https://doi.org/10.3991/ijoe.v7iS2.1769.

García-Zubía, J., Romero, S., Cuadros, J., Guenaga, M., Orduña, P., Dziabenko, O., … Hernández-Jayo, U. (2014). Experiencia de Uso y Evaluación de VISIR en Electrónica Analógica. In *XI Technologies Applied to Electronics Teaching Conference (TAEE)*, Salamanca. IEEE.

García-Zubía, J., Cuadros, J., Serrano, V., Hernández-Jayo, U., Angulo, I., & Villar, A. (2019a). Dashboard for the VISIR remote lab. In *Proceedings of the 5th Experiment at International Conference, exp.at* (pp. 42–46), Funchal, Madeira, Portugal. IEEE. https://doi.org/10.1109/EXPAT.2019.8876527.

García-Zubía, J., Alves, G. R., Hernández-Jayo, U., Cuadros, J., Serrano, V., & Fidalgo, A. (2019b). A framework for interpreting experimental errors in VISIR. In *Proceedings of the 5th Experiment at International Conference, exp.at* (pp. 31–35). Funchal, Madeira, Portugal. IEEE. https://doi.org/10.1109/EXPAT.2019.8876568.

Gustavsson, I., Nilsson, K., Zackrisson, J., García-Zubía, J., Hernández-Jayo, U., Nafalski, A., … Håkansson, L. (2009). On objectives of instructional laboratories, individual assessment, and use of collaborative remote laboratories. *IEEE Transactions on Learning Technologies*, 2(4), 263–274. https://doi.org/10.1109/TLT.2009.42.

Lima, E. (2018). Implantação de Laboratório Remoto em Disciplinas do Curso de Engenharia Elétrica. Universidade Católica de Petrópolis.

Lima, N. (2020). Fostering experimental competences using complementary resources. Universidad de Salamanca. https://doi.org/10.14201/gredos.144127.

Lima, N., Viegas, C., & García-Peñalvo, F. J. (2017a). Learning from Complementary Ways of Experimental Competences Developing Aprendizaje a partir de maneras complementarias de desarrollar capacidades experimentales 1. Context and motivation that drives the dissertation research. *Education in the Knowledge Society (EKS)*, 18(1), 63–74. https://doi.org/10.14201/eks20171816374.

Lima, N., Alves, G. R., Viegas, C., & Gustavsson, I. (2016a). Combined efforts to develop students' experimental competences. In *Proceedings of the 3rd Experiment@ International Conference (exp.at)* (pp. 243–248). https://doi.org/10.1109/EXPAT.2015.7463273.

Lima, N., Viegas, C., Alves, G. R., & García-Peñalvo, F. J. (2016b). VISIR's usage as an educational resource: A review of the empirical research. In *4th International Conference on Technological Ecosystems for Enhancing Multiculturality (TEEM)* (pp. 893–901), New York, NY, USA. ACM. https://doi.org/10.1145/3012430.3012623.

Lima, N., Viegas, C., Alves, G. R., & García-Peñalvo, F. J. (2016c). A utilização do VISIR como um recurso educativo: uma revisão da literatura. In A. Lago Ferreira & G. G. Manuel (Eds.), *TICAI (2016) TICs para el Aprendizaje de la Ingeniería* (pp. 105–114). IEEE, Sociedad de Educación, Capítulos Español y Portugués.

Lima, N., Viegas, C., Alves, G. R., Marques, A., & Fidalgo, A. V. (2021). Students' perception about using VISIR. In *Proceedings of 2021 World Engineering Education Forum/Global Engineering Deans Council, WEEF/GEDC 2021* (pp. 578–583), Madrid, Spain. IEEE. https://doi.org/10.1109/WEEF/GEDC53299.2021.9657418.

Lima, N., Viegas, C., Zannin, M., Marques, A., Alves, G. R., Marchisio, S., … García-Peñalvo, F. J. (2017b). Do students really understand the difference between simulation and remote labs? In *5th International Conference on Technological Ecosystems for Enhancing Multiculturality (TEEM)*, Cádiz, Spain. ACM Press. https://doi.org/10.1145/12345.67890.

Ma, J. & Nickerson, J. (2006). Hands-on, simulated, and remote laboratories: A comparative literature review. *Computer Surveys*, 38(3), Art. no. 7. https://doi.org/10.1145/1132960.1132961.

Marchisio, S., Crepaldo, D., Del Colle, F., Lerro, F. G., Concari, S. B., Leon, D., … Alves, G. R. (2018). VISIR lab integration in electronic engineering: An institutional experience in Argentina. In *2018 XIII Technologies Applied to Electronics Teaching Conference (TAEE)*, Canary Islands, Spain. IEEE. https://doi.org/10.1109/TAEE.2018.8476079.

Marques, A., Viegas, C., Costa-Lobo, M. C., Fidalgo, A., Alves, G. R., Rocha, J. S., & Gustavsson, I. (2014). How remote labs impact on course outcomes: Various practices using VISIR. *IEEE Transactions on Education*, 57(3), 151–159. https://doi.org/10.1109/TE.2013.2284156.

Mendonça, L. N., Maçaneiro, M., Alves, G. R., Pires, D. S., García-Zubía, J., Cuadros, J., & Serrano, V. (2020). Classification of experimental errors done in VISIR with simple alternated current circuits. In *Proceedings of the IEEE Global Engineering Education Conference (EDUCON)* (Vol. 2020-April, pp. 1568–1572), Dubai, UAE. IEEE. https://doi.org/10.1109/EDUCON45650.2020.9125340.

Odeh, S., Alves, G. R., Anabtawi, M., Jazi, M., & Arekat, M. (2014). Experiences with Deploying VISIR at Al-Quds University in Jerusalem. In *Proceedings of the IEEE Global Engineering Education Conference (EDUCON)* (pp. 273–279), Istanbul, Turkey. IEEE.

Pavani, A. M., Lima, D. A., Temporão, G. P., & Alves, G. R. (2018). Different uses for remote labs in electrical engineering education: Initial conclusions of an ongoing experience. In M. Auer & T. Tsiatsos (Eds.), *Advances in Intelligent Systems and Computing* (Vol. 725), Rio de Janeiro, Brazil. https://doi.org/10.1007/978-3-319-75175-7_86.

Pozzo, M. I. (2019). Achievements and challenges of international academic cooperation. The case of VISIR+ project. *IEEE Revista Iberoamericana de Tecnologias del Aprendizaje*, 14(4), 145–151. https://doi.org/10.1109/RITA.2019.2952272.

Pozzo, M. I., Dobboletta, E., Viegas, C., Marques, A., Lima, N., Evangelista, I., & Alves, G. R. (2017). Diseño de instrumentos para la investigación sobre la implementación educativa del laboratorio remoto VISIR en Latinoamérica. In *1er Congresso Latinoamericano de Ingeniería (CLADI)*, Paraná, Entre Ríos, Argentina, 13–15 September 2017.

Romero, S. (2015). AAAS: Modelo de Evaluación Automática de Competencias en el Laboratorio Remoto VISIR, a través de Learning Analytics y Rúbricas de Aprendizaje. PhD Thesis. University of Deusto.

Romero, S., Guenaga, M., García-Zubía, J., & Orduña, P. (2015). Automatic assessment of progress using remote laboratories. *International Journal of Online Engineering*, 11(2), 49–54. https://doi.org/10.3991/ijoe.v11i2.4379.

Sasdelli, I., Alves, G. R., Junior, W. V., & Schlichting, L. C. M. (2021). Caracterização de erros experimentais em circuitos eletrônicos no Laboratório Remoto VISIR+. In *Educação Contemporânea — Volume 15 — Ensino Superior*. Editora Poisson. https://doi.org/10.36229/978-65-5866-057-6.CAP.23.

Serrano, V., Cuadros, J., Fernández-Ruano, L., García-Zubía, J., Hernández-Jayo, U., & Lluch, F. (2022). Learning analytics dashboards for assessing remote labs users' work. A case study with VISIR-DB. In press.

Soysal, O. (2000). Computer integrated experimentation in electrical engineering education over distance. In *Proceedings of the ASEE Annual Conference* (pp. 5.614.1–5.614.10), St. Louis, MO, USA.

Viegas, C., Lima, N., Alves, G. R., & Gustavsson, I. (2014). Improving students' experimental competences using simultaneous methods in class and in assessments. In *2nd International Conference on Technological Ecosystems for Enhancing Multiculturality (TEEM)* (pp. 125–132), Salamanca, Spain. ACM. https://doi.org/10.1145/2669711.2669890.

Viegas, C., Marques, A., Alves, G. R., Alberto, A. S., Dias, C. P., Alves, M. J., & Guimarães, P. S. (2011). On the use of VISIR under different course implementations. In *Proceedings of the 1st Experiment@ International Conference: Online Experimentation (exp.at'11)*, Lisbon, Portugal. Retrieved from http://hdl.handle.net/10400.22/9155.

Viegas, C., Alves, G. R., Marques, A., Lima, N., Felgueiras, M. C., Costa, R. J., … Kreiter, C. (2017). The VISIR+ project — Preliminary results of the training actions. In *Proceedings of the 14th International Conference on Remote Engineering and Virtual Instrumentation (REV)* (Vol. 22, pp. 375–391), New York, USA. https://doi.org/10.1007/978-3-319-64352-6_36.

Viegas, C., Pavani, A. M., Lima, N., Marques, A., Pozzo, M. I., Dobboletta, E., … Alves, G. R. (2018). Impact of a remote lab on teaching practices and student learning. *Computers & Education*, 126, 201–216. https://doi.org/10.1016/j.compedu.2018.07.012.

Chapter 6

The Road Ahead: To Infinity and Beyond

6.1 Introduction

The vision of the VISIR federation, as expressed in Salah (2017), is as follows:

> "*One experiment for all students*
> *All experiments for one student*"

Questioning this vision, what exactly does "all students" mean? And, what about "all experiments"?

The first question links to how many potential users can benefit from VISIR, at any given moment. Considering that VISIR makes it possible to perform experiments with basic-to-complex electrical and electronic circuits, there is a broad range of potential users, including:

- upper secondary school teachers and students, teaching and learning about physics (electricity) (Claesson, 2014),
- teachers and students engaged in undergraduate science and engineering courses that include these sorts of experiments.

According to Statista (2020), in 2019 there were 1.024 million students enrolled in electronics and electrical engineering degrees in India. Given that India awarded 25% of first university degree awards, broadly equivalent to a bachelor's degree, in S&E fields, in global terms, as indicated in Science & Engineering Indicators 2018 (National Science Board, 2018), a simple estimate gives 4.1 million undergraduate students enrolled in electronics and electrical engineering degrees, around the globe. We consider this initial indicator to be the most relevant one, compared with IF (Intensity–Frequency) dimensions, i.e., these students are likely to perform many experiments that can be implemented in VISIR. Students enrolled in upper secondary schools and in Vocational Education Training (VET) programs are likely to perform fewer experiments (frequency↓) but, on the other hand, numbers scale up (intensity↑). According to OECD (2019), on average, around 50% of students that conclude upper secondary education will enter tertiary education. This means the number of students taking Physics in upper secondary schools (in both general and vocational training programs) is in the range of several hundred thousand. Students enrolled in vocational training programs related to electrical and electronic disciplines, however, are likely to benefit more from VISIR, so, for this group, the frequency dimension may be higher.

In addition, Marques *et al.* (2014) provide evidence of VISIR being used in other engineering degrees, such as mechanical and computer engineering. Considering a full semester (15 weeks), VISIR was used for 14 weeks in mechanics and 6 weeks in computer sciences. Again, this would represent a scenario of increased intensity, i.e., according to Statista (2020), the sum of students enrolled in electrical and electronic degrees is 60% smaller when compared to the sum of students enrolled in both mechanical and computer science engineering, and slightly reduced frequency. Furthermore, besides traditional educational contexts (i.e., face-to-face teaching and learning inside formal educational institutions), VISIR has also been used in non-traditional, i.e., informal, and non-formal contexts, for instance, through Massive Open Online Courses (MOOCs), as indicated in Blázquez-Merino *et al.* (2018).

Thus, in conclusion, several hundred thousand students may perform experiments in VISIR every year.

As for the second question, i.e., "how many experiments", this book already presents a first order of magnitude in Part 2, i.e., considering the range of experiments with DC and AC circuits, diodes, transistors, and operational amplifiers, an initial estimation would give slightly over 30 experiments. However, this is an underestimated value simply because the range of possible experiments with both transistors and operational amplifiers, and even with a combination of both, is likely to be in the range of several tenths. Another aspect is the range of possible values for different resistors, capacitors, and inductors (not to mention electronic components). Figure 6.1 shows a typical storage room associated with traditional laboratories for experiments with electrical and electronic circuits. If we consider the number of drawers (in the figure), and that each drawer may contain up to 4 divisions with one component type per division, then a simple calculus of 19 (vertical count) × 33 (horizontal count) gives 627 drawers, which means several hundred components.

Even considering a VISIR matrix with 10 boards, the number of components that can be accommodated is typically in the range of no more

Figure 6.1. A typical storage room associated with traditional laboratories for experiments with electrical and electronic circuits. Top: Piles of 9-drawer electrical and electronic component storage cabinets. Bottom: Detail on drawers with 1, 2, and 4 divisions.

than a few tenths simply because one must also consider the number of shortcuts that need to be installed to measure the current in circuit branches and to allow for different circuit topologies. Again, therefore, in conclusion, it takes more than one single VISIR system to support all possible experiments with electrical and electronic circuits.

6.2 Moving Towards the VISIR Federation

The previous section described one motivation behind the implementation of the VISIR federation, i.e., serving all possible experiments with electrical and electronic circuits to all potential students. However, there is a second motivation based on a model named DIKAR, i.e., Data–Information–Knowledge–Action–Results. Basically, the DIKAR model, as proposed by Venkatraman (1996), provides a framework for achieving results by monitoring the relationship between data, information, knowledge, and strategic actions. In the case of the VISIR federation, the raw data correspond to every experiment performed, i.e., **what** the experiment (circuit topology, components used, instruments configuration and readings, etc.) was; **who** performed the experiment (although these data may be anonymized, there is a unique identifier that may be used to track all experiments made by the same individual); **when** it was performed; and **how long** it took. As for **why** it was performed, the answer may require moving up one step to the information level, i.e., it may require understanding the didactic implementation associated with the experiment performed and the agent.

In sum, having access to all the XML (eXtensible Markup Language) files exchanged with the experiment server of every VISIR node, in addition to the <components.list> and <maxlist> files, provides **data** to obtain **information** on how every VISIR node is being used. Note that ongoing works like Cuadros *et al.* (2021) and Hernández-Jayo *et al.* (2023) already provide tools for extracting information on how many different circuits are being experimented with on a VISIR system and how many correspond to correct or incorrect experiments. This information is important because it helps understand how well VISIR is serving students in their learning process. Performing an incorrect experiment is not necessarily a bad thing if it helps a student discover the path towards performing it correctly. Of course, there is an associated cost (time spent) but, again, there is also an associated gain (learning).

Gathering this information allows the VISIR federation to build **knowledge**, supporting future **actions** and leading to desired **results**. For instance, how close to saturation is one VISIR node? Should a given experiment be replicated in two or more VISIR nodes to distribute the server-access load among those nodes? Should a VISIR node focus on one particular range of experiments to optimize its <component.list>, avoiding unnecessary matrix configurations, or should it try to support a wide range of experiments, in an attempt to support all potential experiments required by the teachers/students using that node?

6.3 The VISIR Roadmap

An alternative application of the DIKAR model, proposed by de Vos (2009) and named RAKID (Results, Actions, Knowledge, Information, Data), makes it possible first to look into a given **result** and then follow the model in the inverse direction, i.e., what **action(s)** should be taken to obtain that result, what **knowledge** is needed to carry out those actions, what **information** is necessary to form that knowledge, and, finally, what sort of **data** are needed to build that information. In fact, this alternative approach has been followed in Alves *et al.* (2022), which describes a roadmap (i.e., a desired result) for VISIR. Quoting the authors:

> The proposed roadmap and its guidelines establish a sustainable strategy and framework to support the future of VISIR and enlarging its community.

The roadmap was defined following a SWOT (Strengths, Weaknesses, Opportunities, and Threats) analysis performed by 15 experts, with extensive experience in VISIR deployment, and their perceptions in three categories: technical, pedagogical, and educational. In each of these categories, future directions have been defined to tackle VISIR weaknesses while not compromising any of its strengths and considering possible opportunities and threats. An aspect worth mentioning is that some of the proposed future directions are in fact already being pursued by ongoing research, such as that of Cosic (2021) and Larbaoui, Naddami, and Fahli (2021). While new researchers are starting their PhDs, addressing specific aspects of VISIR that have been identified in the proposed roadmap, the goal will inevitably be to have the VISIR federation supporting "All experiments for all students", in electrical and electronic circuits.

6.4 An Open Conclusion

In an article entitled "Virtual Laboratories — A historical review and bibliometric analysis of the past three decades", Raman *et al.* (2022) list the top contributing institutions based on publications and citations (referring to articles about virtual laboratories) — see Table 3, page 10 in this volume. Among the top-10 institutions included in that list, six (including the top 2) institutions have a VISIR system installed on their premises and use it with their students. The paper presents further evidence of the relevance of VISIR, including the most cited paper "VISIR: Experiences and challenges" by Javier García-Zubía, who has authored the most publications on the subject (see Table 4, page 12). Table 4 lists the top authors based on publications, citations, and Altmetrics. Taking the column ordered according to the number of total publications, the top 4 authors work with VISIR.

The previous paragraph provides evidence of how VISIR has become a widely disseminated and successful example in terms of remote laboratories.[1] Presently, remote laboratories have attracted the attention of many educational institutions and stakeholders because of the COVID-19 pandemic. In fact, and considering a general audience, speaking of remote laboratories before and after the COVID-19 pandemic represents two entirely different scenarios. The number of emergency remote teaching responses reported in the literature that include the use of remote and virtual laboratories is a simple and undeniable sign. Another piece of evidence, which connects to VISIR, comes from Pablo Orduña, co-founder and CEO of LabsLand,

> The usage of LabsLand remote laboratories has increased substantially since the beginning of the pandemic. In 2020, both the number of sessions and users was 7 times higher, and it is maintaining the growing trend in 2021 (Personal communication, November 5, 2021).

---

[1] Although Raman *et al.* (2022) use the expression "Virtual Laboratories" in the paper's title, a closer look into the keywords used in their query reveals the following: Virtual lab*, online lab*, **remote lab***, virtual experiment*, online experiment*, **remote experiment***, UN SDG*, COVID-19*, and higher education.

LabsLand is a spin-off company of the University of Deusto (Spain), which offers access to VISIR remote laboratories installed at different locations and has the potential to serve as a provider of the VISIR federation. This company has opened access to its remote laboratories, including VISIR, during a large part of the lockdown caused by the COVID-19 pandemic. The growth observed means that many teachers had the opportunity to resort to VISIR to allow their students to perform remote (real) experiments with electrical and electronic circuits. In other words, the lockdown caused by the COVID-19 pandemic created an opportunity for a wider dissemination of the educational value of remote laboratories (and VISIR). In the words (inspired by Max Planck) of Ingvar Gustavsson, creator of VISIR:

> Experimenting could be compared to a conversation with nature. The experimenter asks and Nature answers. The tricky thing is formulating a useful question and above all interpreting the answer. The only way to learn the language of nature is performing many experiments in laboratories that can be hands-on or remote.

Ingvar's words were written before the COVID-19 pandemic. Yet, despite all the evidence reported in Chapter 5 on the educational value of VISIR, there is no doubt the pandemic enlarged the number of teachers, students, and many other stakeholders in education who now understand, and hopefully endorse, Ingvar's words.

References

Alves, G. R., Marques, M. A., Fidalgo, A. V., García-Zubía, J., Castro, M., Hernández-Jayo, U., García-Loro, F., & Kreiter, C. (2022). A roadmap for the VISIR Remote Lab. *European Journal of Engineering Education*. https://doi.org/10.1080/03043797.2022.2054312.

Alves, G. R., Pester, A., Kulesza, W. J., Silva, J. B., Pavani, A. M., Pozzo, M. I., … García-Zubía, J. (2018). A sustainable approach to let students perform more real experiments with electrical and electronic circuits. In *6th International Conference on Technological Ecosystems for Enhancing Multiculturality (TEEM)* (pp. 508–514), Salamanca, Spain. ACM Press. https://doi.org/10.1145/3284179.3284265.

Blázquez-Merino, M., Macho-Aroca, A., Baizán-Álvarez, P., García-Loro, F., Sancristobal, E., Díaz-Orueta, G., & Castro, M. (2018). Structured MOOC designed to optimize electricity learning at secondary school. In *Proceedings of the IEEE Global Engineering Education Conference (EDUCON)* (pp. 223–232), Canary Islands, Spain. IEEE. https://doi.org/10.1109/EDUCON.2018.8363232.

Claesson, L. (2014). Remote Electronic and Acoustic Laboratories in Upper Secondary Schools. Licentiate dissertation. Blekinge Institute of Technology. Retrieved from http://urn.kb.se/resolve?urn=urn:nbn:se:bth-00593.

Cosic, D. (2021). VISIR Remote Lab controlled via VR. Carinthia University of Applied Sciences. (CUAS), Austria. Master Thesis.

Cuadros, J., Serrano, V., Lluch, F., García-Zubía, J., & Hernández-Jayo, U. (2021). Mapping VISIR circuits for computer-assisted assessment. In *Proceedings of 2021 World Engineering Education Forum/Global Engineering Deans Council, WEEF/GEDC 2021* (pp. 524–527), Madrid, Spain. IEEE. https://doi.org/10.1109/WEEF/GEDC53299.2021.9657349.

de Vos, K. (producer) (2009). Knowledge management. Retrieved from https://www.slideshare.net/deVos/general-knowledge-management-overview.

Hernández-Jayo, U., García-Zubía, J., Cuadros, J., Serrano, V., Fernandez-Ruano, L. & Alves, G. R. (2023). Automatic Assessment using VISIR-DB. *Proceedings of the 20th International Conference on Remote Engineering and Virtual Instrumentation*, Thessaloniki, Greece, 1–3 March, 2023.

Larbaoui, Y., Naddami, A., & Fahli, A. (2021). Switching matrix architecture for flexible remote experiments of circuits structuring in electronics and electricity while using an intelligent algorithm. *International Journal of Scientific & Technology Research*, 10(2), 306–313.

Marques, A., Viegas, C., Costa-Lobo, M. C., Fidalgo, A., Alves, G. R., Rocha, J. S., & Gustavsson, I. (2014). How remote labs impact on course outcomes: Various practices using VISIR. *IEEE Transactions on Education*, 57(3), 151–159. https://doi.org/10.1109/TE.2013.2284156.

National Science Board (2018). Science & Engineering Indicators 2018. https://www.nsf.gov/statistics/2018/nsb20181/assets/nsb20181.pdf.

OECD (2019). *Education at a Glance 2019: OECD Indicators*. Paris: OECD Publishing. https://doi.org/10.1787/f8d7880d-en.

Raman, R., Achuthan, K., Nair, V. K., *et al.* (2022). Virtual laboratories — A historical review and bibliometric analysis of the past three decades. *Education and Information Technologies*. https://doi.org/10.1007/s10639-022-11058-9.

Salah, R. M. (2017). A federation of online labs for assisting science and engineering education in the MENA region. Faculty of Science and Technology of the University of Algarve. Retrieved from http://hdl.handle.net/10400.1/10100.

Statista (2020). Number of undergraduate engineering students in India 2019 by discipline. Published by Statista Research Department, October 16, 2020. https://www.statista.com/statistics/765482/india-number-of-students-enrolled-in-engineering-stream-by-discipline/.

Venkatraman, N. (1996). Managing IT resources as a value center. IS Executive Seminar Series, Cranfield School of Management.

Index

www.ingramcontent.com/pod-product-compliance
Lightning Source LLC
Chambersburg PA
CBHW050549190326
41458CB00007B/1976

* 9 7 8 9 8 1 1 2 7 4 1 4 5 *